Anonymous

Studies in Biology

Vol. 3

Anonymous

Studies in Biology
Vol. 3

ISBN/EAN: 9783337219901

Printed in Europe, USA, Canada, Australia, Japan

Cover: Foto ©berggeist007 / pixelio.de

More available books at **www.hansebooks.com**

STUDIES

FROM THE

BIOLOGICAL DEPARTMENT.

STUDIES

IN

BIOLOGY

FROM THE

BIOLOGICAL DEPARTMENT

OF

THE OWENS COLLEGE.

VOLUME III.

PUBLISHED BY THE COUNCIL OF THE COLLEGE

AND EDITED BY

PROFESSOR SYDNEY J. HICKSON.

MANCHESTER:
J. E. CORNISH
1895.

PRICE TEN SHILLINGS

PREFACE.

The present volume of the Studies contains a series of papers that were either written or published before I entered upon my duties as Beyer Professor of Zoology, and, with the exception of the essay by Dr. Hurst on the Structure of Archaeopteryx, they were all published during the lifetime of my distinguished predecessor, the late Professor Arthur Milnes Marshall.

I cannot claim therefore that any of the work recorded in this volume was done at my advice or under my guidance and supervision, nor can I, on the other hand, be held responsible for the accuracy of the observations or for the opinions expressed in their essays by their several authors.

The delay that has occurred in the publication of the volume, consequent on the sudden death of Dr. Marshall and the lapse of time before appointment of his successor, was unavoidable; but I trust that the appearance of the IIIrd volume of the Studies will be none the less welcome to those who are interested in the work that is being carried on in the scientific laboratories of our College.

I would take this opportunity of reminding those Zoologists who read this volume that on the death of Dr. Marshall his relatives very generously gave his scientific books and pamphlets to the Owens College, for the use of the students and teachers of the same for all time. It would add very greatly to the value of this collection if Zoologists would kindly forward to me, for inclusion in 'the Marshall Library,' any separate copies of their publications that they can spare. Such donations would be most gratefully received, as certain to prove of very considerable assistance to those of us who are engaged in original work in this laboratory.

The third volume of the Studies, although it is the last in which any part was taken by Professor Marshall, does not, by any means, record the end of his work. The stimulus of his teaching and the energy of his character must remain with us in the Owens College for many years to come, and his influence will long be felt in the writings both of his pupils and successors.

SYDNEY J. HICKSON.

September, 1895.

CONTENTS.

ADDRESS TO THE BIOLOGICAL SECTION OF THE BRITISH ASSOCIATION (1890),

By Professor A. MILNES MARSHALL, M.A., M.D., D.Sc., F.R.S.,
President of the Section.

As my theme for this morning's address I have selected the Development
of Animals. I have made this choice from no desire to extol one
particular branch of biological study at the expense of others, nor
through failure to appreciate or at least admire the work done and the
results achieved in recent years by those who are attacking the great
problems of life from other sides and with other weapons.

My choice is determined by the necessity that is laid upon me,
through the wide range of sciences whose encouragement and advancement are the peculiar privilege of this Section, to keep within reasonable
limits the direction and scope of my remarks ; and is confirmed by the
thought that, in addressing those specially interested in and conversant
with biological study, your President acts wisely in selecting as the
subject-matter of his discourse some branch with which his own studies
and inclinations have brought him into close relation.

Embryology, referred to by the greatest of naturalists as 'one of the
most important subjects in the whole round of Natural History,' is still
in its youth, but has of late years thriven so mightily that fear has
been expressed lest it should absorb unduly the attention of zoologists
or even check the progress of science by diverting interest from other
and equally important branches.

Nor is the reason of this phenomenal success hard to find. The

actual study of the processes of development; the gradual building up
of the embryo, and then of the young animal, within the egg; the
fashioning of its various parts and organs; the devices for supplying
it with food, and for ensuring that the respiratory and other interchanges
are duly performed at all stages: all these are matters of absorbing
interest. Add to these the extraordinary changes which may take
place after leaving the egg, the conversion, for instance, of the aquatic
gill-breathing tadpole—a true fish as regards all essential points of its
anatomy—into a four-legged frog, devoid of tail, and breathing by lungs;
or the history of the metamorphosis by which the sea-urchin is gradually
built up within the body of its pelagic larva, or the butterfly derived
from its grub. Add to these again the far wider interest aroused by comparing the life histories of allied animals, or by tracing the mode of
development of a complicated organ, *e.g.* the eye or the brain, in the
in the various animal groups, from its simplest commencement, through
gradually increasing grades of efficiency, up to its most perfect form as
seen in the highest animals. Consider this, and it becomes easy to
understand the fascination which embryology exercise over those who
study it.

But all this is of trifling moment compared with the great generali-
sation which tells us that the development of animals has a far higher
meaning; that the several embryological stages and the order of their
occurrence are no mere accidents, but are forced on an animal in
accordance with a law, the determination of which ranks as one of the
greatest achievements of biological science.

The doctrine of descent, or of Evolution, teaches us that as
individual animals arise, not spontaneously, but by direct descent from
pre-existing animals, so also is it with species, with families, and with
larger groups of animals, and so also has it been for all time; that as
the animals of succeeding generations are related together, so also are
those of successive geologic periods; that all animals, living or that
have lived, are united together by blood relationship of varying near-
ness or remoteness; and that every animal now in existence has a
pedigree stretching back, not merely for ten or a hundred generations,
but through all geologic time since the dawn of life on this globe.

The study of Development, in its turn, has revealed to us that each
animal bears the mark of its ancestry, and is compelled to discover its
parentage in its own development; that the phases through which an

animal passes in its progess from the egg to the adult are no accidental freaks, no mere matters of developmental convenience, but represent more or less closely, in more or less modified manner, the ancestral stages through which the present condition has been acquired.

Evolution tells us that each animal has had a pedigree in the past. Embryology reveals to us this ancestry, because every animal in its own development repeats this history, climbs up its own genealogical tree.

Such is the Recapitulation Theory, hinted at by Agassiz, and suggested more directly in the writings of von Baer, but first clearly enunciated by Fritz Müller, and since elaborated by many, notably by Balfour and by Ernst Haeckel.

It is concerning this theory, which forms the basis of the Science of Embryology, and which alone justifies the extraordinary attention this science has received, that I venture to address you this morning.

A few illustrations from different groups of animals will best explain the practical bearings of the theory, and the aid which it affords to the zoologist of to-day ; while these will also serve to illustrate certain of the difficulties which have arisen in the attempt to interpret individual development by the light of past history—difficulties which I propose to consider at greater length.

A very simple example of recapitulation is afforded by the eyes of the sole, plaice, turbot, and their allies. These 'flat fish' have their bodies greatly compressed laterally ; and the two surfaces, really the right and left sides of the animal, unlike, one being white, or nearly so, and the other coloured. The flat fish has two eyes, but these, in place of being situated, as in other fish, one on each side of the head, are both on the coloured side. The advantage to the fish is clear, for the natural position of rest of a flat fish is lying on the sea bottom, with the white surface downwards and the coloured one upwards. In such a position an eye situated on the white surface could be of no use to the fish, and might even become a source of danger, owing to its liability to injury from stones or other hard bodies on the sea bottom.

No one would maintain that flat fish were specially created as such. The totality of their organisation shows clearly enough that they are true fish, akin to others in which the eyes are symmetrically placed one on each side of the head, in the position they normally hold among vertebrates. We must therefore suppose that flat fish are descended from other fish in which the eyes are normally situated.

The Recapitulation Theory supplies a ready test. On employing it, *i.e.*, on studying the development of the flat fish, we obtain a conclusive answer. The young sole on leaving the egg is shaped just as any ordinary fish, and has the two eyes placed symmetrically on the two sides of the head. It is only after the young fish has reached some size, and has begun to approach the adult in shape, and to adopt its habit of resting on one side on the sea bottom, that the eye of the side on which it rests becomes shifted forwards, then rotated on to the top of the head, and finally twisted completely over to the opposite side.

The brain of a bird differs from that of other vertebrates in the position of the optic lobes, these being situated at the sides instead of on the dorsal surface. Development shows that this lateral position is a secondarily acquired one, for throughout all the earlier stages the optic lobes are, as in other vertebrates, on the dorsal surface, and only shift down to the sides shortly before the time of hatching.

Crabs differ markedly from their allies, the lobsters, in the small size and rudimentary condition of their abdomen or "tail." Development, however, affords abundant evidence of the descent of crabs from macrurous ancestors, for a young crab at what is termed the Megalopa stage has the abdomen as large as a lobster or prawn at the same stage.

Molluscs afford excellent illustrations of recapitulation. The typical gastropod has a large spirally-coiled shell ; the limpet, however, has a large conical shell, which in the adult gives no sign of spiral twisting, although the structure of the animal shows clearly its affinity to forms with spiral shells. Development solves the riddle at once, telling us that in its early stages the limpet embryo has a spiral shell, which is lost on the formation, subsequently, of the conical shell of the adult.

Recapitulation is not confined to the higher groups of animals, and the Protozoa themselves yield most instructive examples. A very striking case is that of Orbitolites, one of the most complex of the porcellanous Foraminifera, in which each individual during its own growth and development passes through the series of stages by which the cyclical or discoidal type of shell was derived from the simpler spiral form.

In *Orbitolites tenuissima*, as Dr. Carpenter has shown,[1] 'the whole

[1] W. B. Carpenter, 'On an Abyssal Type of the Genus Orbitolites,' *Phil. Trans.* 1883, part ii. p. 553.

transition is actually presented during the successive stages of its growth. For it begins life as a Cornuspira, ... its shell forming a continuous spiral tube, with slight interruptions at the points at which its successive extensions commence ; while its sarcolic body is coiled into a continuous coil with slight constrictions at intervals. The second stage consists in the opening out of its spire, and the division of its cavity at regular intervals by transverse septa, traversed by separate pores, exactly as in Peneroplis. The third stage is marked by the subdivision of the "peneropline" chambers into chamberlets, as in the early forms of Orbiculina. And the fourth consists in the exchange of the spiral for the cyclical plan of growth, which is characteristic of Orbitolites ; a circular disc of progressively increasing diameter being formed by the addition of successive annular zones around the entire periphery.'

The shells both of Foraminifera and of Mollusca afford peculiarly instructive examples for the study of recapitulation. As growth of the shell is effected by the addition of new shelly matter to the part already existing, the older parts of the shell are retained, often unaltered, in the adult ; and in favourable cases, as in *Orbitolites tenuissima*, all the stages of development can be determined by simple inspection of the adult shell.

It is important to remember that the Recapitulation Theory, if valid, must apply not merely in a general way to the development of the animal body, but must hold good with regard to the formation of each organ or system, and with regard to the later equally with the earlier phases of development.

Of individual organs the brain of birds has been already cited. The formation of the vertebrate liver as a diverticulum from the alimentary canal, which is at first simple, but by the folding of its walls becomes greatly complicated, is another good example ; as is also the development of the vomer in Amphibians as a series of toothed plates, equivalent morphologically to the placoid scales of fishes, which are at first separate, but later on fuse together and lose the greater number of their teeth.

Concerning recapitulation in the latter phases of development and in the adult animal, the mode of renewal of the nails or of the epidermis generally is a good example, each cell commencing its existence in an indifferent form in the deeper layers of the epidermis, and gradually acquiring the adult peculiarities as it approaches the surface, through removal of the cells lying above it.

The above examples, selected almost haphazard, will suffice to illustrate the Theory of Recapitulation.

The proof of the theory depends chiefly on its universal applicability to all animals, whether high or low in the zoological scale, and to all their parts and organs. It derives also strong support from the ready explanation which it gives of many otherwise unintelligible points.

Of these latter a familiar and most instructive instance is afforded by rudimentary organs, *i.e.*, structures which, like the outer digits of the horse's leg, or the intrinsic muscles of the ear of a man, are present in the adult in an incompletely developed form, and in a condition in which they can be of no use to their possessors; or else structures which are present in the embryo, but disappear completely before the adult condition is attained, for example, the teeth of whalebone whales, or the branchial clefts of all higher vertebrates.

Natural selection explains the preservation of useful variations, but will not account for the formation and perpetuation of useless organs; and rudiments such as those mentioned above would be unintelligible but for Recapitulation, which solves the problem at once, showing that these organs, though now useless, must have been of functional value to the ancestors of their present possessors, and that their appearance in the ontogeny of existing forms is due to repetition of ancestral characters. Such rudimentary organs are, as Darwin pointed out, of larger relative or even absolute size in the embryo than in the adult, because the embryo represents the stage in the pedigree in which they were functionally active.

Rudimentary organs are extremely common, especially among the higher groups of animals, and their presence and significance are now well understood. Man himself affords numerous and excellent examples, not merely in his bodily structure, but by his speech, dress, and customs. For the silent letter *b* in the word doubt, or the *w* of answer, or the buttons on his elastic-side boots are as true examples of rudiments, unintelligible but for their past history, as are the ear muscles he possesses but cannot use, or the gill-clefts, which are functional in fishes and tadpoles, and are present, though useless, in the embryos of all higher vertebrates, which in their early stages the hare and the tortoise alike possess, and which are shared with them by cats and by kings.

Another consideration of the greatest importance arises from the study of the fossil remains of the animals that formerly inhabited the earth. It was the elder Agassiz who first directed attention to the remarkable agreement between the embryonic growth of animals and their palæontological history. He pointed out the resemblance between certain stages in the growth of young fish and their fossil representatives, and attempted to establish, with regard to fish, a correspondence between their palæontological sequence and the successive stages of embryonic development. He then extended his observations to other groups, and stated his conclusions in these words: "It may therefore be considered as a general fact, very likely to be more fully illustrated as investigations cover a wider ground, that the phases of development of all living animals correspond to the order of succession of their extinct representatives in past geological times."[1]

This point of view is of the utmost importance. If the development of an animal is really a repetition of its ancestral history, then it is clear that the agreement or parallelism which Agassiz insists on between the embryological and palæontological records must hold good. Owing to the attitude which Agassiz subsequently adopted with regard to the theory of Natural Selection, there is some fear of his services in this respect failing to receive full recognition, and it must not be forgotten that the sentence I have quoted was written prior to the clear enunciation of the Recapitulation Theory by Fritz Müller.

The imperfection of the geological record has been often referred to and lamented. It is very true that our museums afford us but fragmentary pictures of life in past ages; that the earliest volumes of the history are lost, and that of others but a few torn pages remain to us; but the later records are in far more satisfactory condition. The actual number of specimens accumulated from the more recent formations is prodigious; facilities for consulting them are far greater than they were; the international brotherhood of science is now fully established, and the fault will be ours if the material and opportunities now forthcoming are not rightly and fully utilised.

By judicious selection of groups in which long series of specimens can be obtained, and in which the hard skeletal parts, which alone can be suitably preserved as fossils, afford reliable indications of zoological affinity, it is possible to test directly this correspondence

[1] L. Agassiz, *Essay on Classification*, 1859, p. 115.

between palæontological and embryological histories, while in some instances a single lucky specimen will afford us, on a particular point, all the evidence we require.

Great progress has already been made in this direction, and the results obtained are of the most encouraging description.

By Alexander Agassiz a detailed comparison was made between the fossil series and the developmental stages of recent forms in the case of the Echinoids, a group peculiarly well adapted for such an investigation. The two records agree remarkably in many respects, more especially in the independent evidence they give as to the origin of the asymmetrical forms from more regular ancestors. The gradually increasing complication in some of the historic series is found to be repeated very closely in the development of their existing representatives; and with regard to the whole group, Agassiz concludes that,[1] 'comparing the embryonic development with the palæontological one, we find a remarkable similarity in both, and in a general way there seems to be a parallelism in the appearance of the fossil genera and the successive stages of the development of the Echini.'

Neumayr has followed similar lines, and by him, as by other authorities on the group, there seems to be general agreement as to the parallelism between the embryological and palæontological records, not merely for Echini, but for other groups of Echinodermata as well.

The Tetrabranchiate Cephalopoda are an excellent group in which to study the problem, for though no opportunity has yet occurred for studying the embryology of the only surviving member of the group the pearly nautilus, yet owing to the fact that growth of the shell is effected by addition of shelly matter to the part already present, and to the additions being made in such manner that the older part of the shell persists unaltered, it is possible, from examination of a single shell—and in the case of fossils the shells are the only part of which we have exact knowledge—to determine all the phases of its growth ; just as in the shell of Orbitolites all the stages of development are manifest on inspection of an adult specimen.

In such a shell as Nautilus or Ammonites the central chamber is the oldest or first formed one, to which the remaining chambers are added

---

[1] A. Agassiz, *Palæontological and Embryological Development.* 'An Address before the American Association for the Advancement of Science.' 1880.

in succession. If, therefore, the development of the shell is a repetition of ancestral history, the central chamber should represent the palæontologically oldest form, and the remaining chambers in succession forms of more and more recent origin. Ammonite shells present, more especially in their sutures, and in the markings and sculpturing of their surface, characters that are easily recognised, and readily preserved in fossils; and the group, consequently, is a very suitable one for investigation from this standpoint.

Würtenberg's admirable and well-known researches[1] have shown that in the Ammonites such a correspondence between the historic and embryonic development does really exist; that, for example, in Aspidoceras the shape and markings of the shells in young specimens differ greatly from those of adults, and that the characters of the young shells are those of palæontologically older forms.

Another striking illustration of the correspondence between the palæontological and developmental records is afforded by the antlers of deer, in which the gradually increasing complication of the antler in successive years agrees singularly closely with the progressive increase in size and complexity shown by the fossil series from the Miocene age to recent times.

Of cases where a single specimen has sufficed to prove the palæontological significance of a developmental character, Archæopteryx affords a typical example. In recent birds the metacarpals are firmly fused with one another, and with the distal series of carpals; but in development the metacarpals are at first, and for some time, distinct. In Archæopteryx this distinctness is retained in the adult, showing that what is now an embryonic character in recent birds, was formerly an adult one.

Other examples might easily be quoted, but these will suffice to show that the relation between Palæontology and Embryology, first enunciated by Agassiz, and required by the Recapitulation Theory, does in reality exist. There is much yet to be done in this direction. A commencement, a most promising commencement, has been made, but as yet only a few groups have been seriously studied from this standpoint.

It is a great misfortune that palæontology is not more generally

---

[1] L. Würtenberger, 'Studien über die Stammesgeschichte der Ammoniten. Ein geologischer Beweis für die Darwin'sche Theorie.' Leipzig, 1880.

and more seriously studied by men versed in embryology, and that
those who have so greatly advanced our knowledge of the early
development of animals should so seldom have tested their conclusions
as to the affinities of the groups they are concerned with by direct
reference to the ancestors themselves, as known to us through their
fossil remains.

I cannot but feel that, for instance, the determination of the
affinities of fossil Mammalia, of which such an extraordinary number
and variety of forms are now known to us, would be greatly facilitated
by a thorough and exact knowledge of the development, and especially
the later development, of the skeleton in their existing descendants,
and I regard it as a reproach that such exact descriptions of the later
stages of development should not exist even in the case of our
commonest domestic animals.

The pedigree of the horse has attracted great attention, and has
been worked at most assiduously, and we are now, largely owing to the
labours of American palæontologists, able to refer to a series of fossil
forms commencing in the lowest Eocene beds, and extending upwards
to the most recent deposits, which show a complete gradation from a more
generalised mammalian type to the highly specialised condition
characteristic of the horse and its allies, and which may reasonably be
regarded as indicating the actual line of descent of the horse. In this
particular case, more frequently cited than any other, the evidence is
entirely palæontological. The actual development of the horse has yet
to be studied, and it is greatly to be desired that it should be under-
taken speedily. Klever's[1] recent work on the development of the
teeth in the horse may be referred to as showing that important and
unexpected evidence is to be obtained in this way.

A brilliant exception to the statement just made as to the want of
exact knowledge of the later development of the more highly organised
animals is afforded by the splendid labours of Professor Kitchen
Parker, whose recent death has deprived zoology of one of her most
earnest and single-minded students, and zoologists, young and old
alike, of a true and sincere friend. Professor Parker's extraordinarily
minute and painstaking investigations into the development of the
vertebrate skull rank among the most remarkable of modern zootomical

[1] Klever, 'Zur Kenntnis der Morphogenese des Equidengebisses,'
Morphologisches Jahrbuch xv. 1889, p. 308.

achievements, and afford a rich mine of carefully recorded facts, the full value and bearing of which we are hardly yet able to appreciate.

If further evidence as to the value and importance of the Recapitulation Theory were needed, it would suffice to refer to the influence which it has had on the classification of the animal Kingdom. Ascidians and Cirripedes may be quoted as important groups, the true affinities of which were first revealed by embryology; and in the case of parasitic animals the structural modifications of the adult are often so great that but for the evidence yielded by development their zoological position could not be determined. It is now indeed generally recognised that in doubtful cases embryology affords the safest of all clues, and that the zoological position of such forms can hardly be regarded as definitely established unless their development, as well as their adult anatomy is ascertained.

It is owing to this Recapitulation Theory that Embryology as exercised so marked an influence on zoological speculation. Thus the formation in most, if not in all, animals of the nervous system and of the sense organs from the epidermal layer of the skin, acquired a new significance when it was recognised that this mode of development was to be regarded as a repetition of the primitive mode of formation of such organs; while the vertebral theory of the skull affords a good example of a view, once stoutly maintained, which received its death-blow through the failure of embryology to supply the evidence requisite in its behalf. The necessary limits of time and space forbid that I should attempt to refer to even the more important of the numerous recent discoveries in embryology, but mention may be very properly made here of Sedgwick's determination of the mode of development of the body cavity in Peripatus, a discovery which has thrown most welcome light on what was previously a great morphological puzzle.

We must now turn to another side of the question. Although it is undoubtedly true that development is to be regarded as a recapitulation of ancestral phases, and that the embryonic history of an animal presents to us a record of the race history, yet it is also an undoubted fact, recognised by all writers on embryology, that the record so obtained is neither a complete nor a straightforward one.

It is indeed a history, but a history of which entire chapters are lost, while in those that remain many pages are misplaced while others are so blurred as to be illegible; words, sentences, or entire paragraphs are omitted, and worse still, alterations or spurious additions have been freely introduced by later hands, and at times so cunningly as to defy detection.

Very slight consideration will show that development cannot in all cases be strictly a recapitulation of ancestral stages. It is well known that closely allied animals may differ markedly in their mode of development. The common frog is at first a tadpole, breathing by gills, a stage which is entirely omitted by the West Indian Hylodes. A cray fish, a lobster, and a prawn are allied animals, yet they leave the egg in totally different forms. Some developmental stages, as the pupa condition of insects, or the stage in the development of a dogfish, in which the œsophagus is imperforate, cannot possibly be ancestral stages. Or again, a chick embryo of, say the fourth day, is clearly not an animal capable of independent existence, and therefore cannot correctly represent any ancestral condition, an objection which applies to the developmental history of many, perhaps of most animals.

Haeckel long ago urged the necessity of distinguishing in actual development between those characters which are really historical and inherited and those which are acquired or spurious additions to the record. The former he termed palingenetic or ancestral characters, the latter cenogenetic or acquired. The distinction is undoubtedly a true one, but an exceedingly difficult one to draw in practice. The causes which prevent development from being a strict recapitulation of ancestral characters, the mode in which these came about, and the influence which they respectively exert, are matters which are greatly exercising embryologists, and the attempt to determine which has as yet met with only partial success.

The most potent and the most widely spread of these disturbing causes arise from the necessity of supplying the embryo with nutriment. This acts in two ways. If the amount of nutritive matter within the egg is small, then the young animal must hatch early, and in a condition in which it is able to obtain food for itself. In such cases there is of necessity a long period of larval life, during which natural selection may act so as to introduce modifications of the ancestral history, spurious additions to the text.

If, on the other hand, the egg contain within itself a considerable quantity of nutrient matter, then the period of hatching can be postponed until this nutrient matter has been used up. The consequence

is that the embryo hatches at a much later stage of its development,
and if the amount of food material is sufficient may even leave the egg
in the form of the parent. In such cases the earlier developmental
phases are often greatly condensed and abbreviated ; and as the
embryo does not lead a free existence, and has no need to exert itself
to obtain food, it commonly happens that these stages are passed
through in a very modified form, the embryo being as in a four-day
chick, in a condition in which it is clearly incapable of independent
existence.

The nutrition of the embryo prior to hatching is most usually
effected by granules of nutrient matter, known as food yolk, and
embedded in the protoplasm of the egg itself ; and it is on the
relative abundance of these granules that the size of the egg chiefly
depends.

Large size of eggs implies diminution of number of the eggs, and
hence of the offspring ; and it can be well understood that while some
species derive advantage in the struggle for existence by producing the
maximum number of young, to others it is of greater importance that
the young on hatching should be of considerable size and strength, and
able to begin the world on their own account. In other words, some
animals may gain by producing a large number of small eggs, others by
producing a smaller number of eggs of larger size—*i.e.*, provided with
more food yolk.

The immediate effect of a large amount of food yolk is to mechanically
retard the processes of development ; the ultimate result is to greatly
shorten the time occupied by development. This apparent paradox is
readily explained. A small egg, such as that of Amphioxus, starts its
development rapidly, and in about eighteen hours gives rise to a free
swimming larva, capable of independent existence, with digestive
cavity and nervous system already formed ; while a large egg like
that of the hen, hampered by the great mass of food yolk by which it
is distended, has, in the same time, made but very slight progress.

From this time, however, other considerations begin to tell. Amphioxus has been able to make this rapid start owing to its relative
freedom from food yolk. This freedom now becomes a retarding
influence, for the larva, containing within itself but a very scanty
supply of nutriment, must devote much of its energies to hunting for
and to digesting its food, and hence its further development will
proceed more slowly.

The chick embryo, on the other hand, has an abundant supply of food in the egg itself; it has no occasion to spend time searching for food, but can devote its whole energies to the further stages of its development. Hence, except in the earliest stages, the chick develops more rapidly than Amphioxus, and attains its adult form in a much shorter time.

The tendency of abundant food yolk to lead to shortening or abbreviation of the ancestral history, and even to the entire omission of important stages, is well known. The embryo of forms well provided with yolk takes short cuts in its development, jumps from branch to branch of its genealogical tree, instead of climbing steadily upwards.

Thus the little West Indian frog, Hylodes, produces eggs which contain a larger amount of food yolk than those of the common English frog. The young Hylodes is consequently enabled to pass through the tadpole stage before hatching, to attain the form of a frog before leaving the egg; and the tadpole stage is only imperfectly recapitulated, the formation of gills, for instance, being entirely omitted.

The influence of food yolk on the development of animals is closely analogous to that of capital in human undertakings. A new industry, for example, that of pen-making, has often been started by a man working by hand and alone, making and selling his own wares; if he succeed in the struggle for existence, it soon becomes necessary for him to call in others to assist him, and to subdivide the work: hand labour is soon superseded by machines, involving further differentiation of labour; the earlier machines are replaced by more perfect and more costly ones; factories are built, agents engaged, and in the end, a whole army of workpeople employed. In later times a man commencing business with very limited means will start at the same level as the original founder, and will have to work his way upwards through much the same stages, i.e., will repeat the pedigree of the industry. The capitalist, on the other hand, is enabled, like Hylodes, to omit these earlier stages, and after a brief period of incubation, to start business with large factories equipped with the most recent appliances, and with a complete staff of workpeople, i.e., to spring into existence fully fledged.

There is no doubt that abundance of food yolk is a direct and very frequent cause of the omission of ancestral stages from individual development; but it must not be viewed as a sole cause. It is quite impossible that any animal, except perhaps in the lowest zoological groups, should repeat all the ancestral stages in the history of the race; the limits of time available for individual development will not permit this. There is a tendency in all animals towards condensation of the ancestral history, towards striking a direct path from the egg to the adult.

This tendency is best marked in the higher, the more complicated members of a group, i.e., in those which have a longer and more tortuous pedigree; and though greatly strengthened by the presence of food yolk in the egg, is apparently not due to this in the first instance.

Thus the simpler forms of Orbitolites, as O. *tenuissima*, repeat in their development all the stages leading from a spiral to a cyclical shell; but in the more complicated species, as Dr. Carpenter has pointed out, there is a tendency towards precocious development of the adult characters, the earlier stages being hurried over in a modified form; while in the most complex examples, as in O. *complanata*, the earlier spiral stages may be entirely omitted, the shell acquiring almost from its earliest commencement the cyclical mode of growth. There is no question here of relative abundance of food yolk, but merely of early or precocious appearance of adult characters.

The question of the relations and influence of food yolk, involving as it does the larger or smaller size of the egg, is, however, merely a special side of the much wider question of the nutrition of the embryo, one of the most potent of the disturbing elements affecting development.

Speaking generally, we may say that large eggs are more often met with in the higher than the lower groups of animals. Birds and Reptiles are cases in point and, if Mammals do not now produce large eggs, it is because a more direct and more efficient mode of nourishing the young by the placenta has been acquired by the higher forms, and has replaced the food-yolk that was formerly present, and is now retained in quantity by Monotremes alone. Molluscs afford another good example, the eggs of Cephalopoda being of larger size than those of the less highly organised groups.

The large size of the eggs of Elasmobranchs, and perhaps that of Cephalopods also, may possibly be associated with the carnivorous habits of the animals; for it is of importance that forms which prey on other animals should hatch of considerable size and strength.

The influence of habitat must also be considered. It has long been noticed as a general rule that marine animals lay small eggs, while their fresh-water allies have eggs of much larger size. The eggs of the salmon or trout are much larger than those of the cod or herring; and the crayfish, though only a quarter the length of a lobster, lay eggs of actually larger size.

This larger size of the eggs of fresh-water forms appears to be dependent on the nature of the environment to which they are exposed. Considering the geological instability of the land as compared with the ocean, there can be no doubt that the fresh-water fauna is, speaking generally, derived from the marine fauna; and the great problem with regard to fresh-water life is to explain why it is that so many groups of animals which flourish abundantly in the sea should have failed to establish themselves in fresh water. Sponges and Cœlenterates abound in the sea, but their fresh-water representatives are extremely few in number; Echinoderms are exclusively marine; there are no fresh-water Cephalopods, and no Ascidians: and of the smaller groups of Worms, Molluscs, and Crustaceans, there are many that do not occur in fresh water.

Direct experiment has shown that in many cases this distribution is not due to inability of the adult animals to live in fresh water; and the real explanation appears to be that the early larval stages are unable to establish themselves under such conditions. This interesting suggestion, which has been worked out in detail by Professor Sollas[1] undoubtedly affords an important clue. To establish itself permanently in fresh water an animal must either be fixed, or else be strong enough to withstand and make headway against the currents of the streams or rivers it inhabits, for otherwise it will in the long run be swept out to sea, and this consideration applies to larval forms equally with adults.

The majority of marine Invertebrates leave the egg as minute ciliated larvæ: and such larvæ are quite incapable of holding their own in currents of any strength. Hence, it is only forms which have

[1] W. J. Sollas, 'On the Origin of Freshwater Faunas.' *Scientific Transactions of the Royal Dublin Society*, vol. iii. Ser. 11, 1886.

got rid of the free swimming ciliated larval stage, and which leave the
egg of considerable size and strength, that can establish themselves as
fresh water animals. This is effected most readily by the acquisition of
food yolk—hence the large size of the eggs of fresh-water animals—and
is often supplemented, as Sollas has shown, by special protective devices
of a most interesting nature. For this reason fresh water forms are not
so well adapted as their marine allies for the study of ancestral history
as revealed in larval or embryonic development.

Before leaving the question of food yolk, reference must be made to
the proposal of the brothers Sarasin, to regard the yolk cells as forming
a distinct embryonic layer, the lecithoblast,[1] distinct from the blastoderm.
I do not desire to speak dogmatically on a point the full bearings of
which are not yet apparent, but I venture to think that this suggestion
will not commend itself to embryologists. The distinction between the
yolk granules and the cells in which they are imbedded is a real
and fundamental one; but I see no reason for regarding the yolk cells
as other than originally functional endoderm cells in which yolk
granules have accumulated to such an extent that they have in extreme
cases become devoted solely to the storing of food for the embryo.[2]

Of all the causes tending to modify development, tending to obscure
or falsify the ancestral record, food yolk is the most frequent and the
most important; its position in the egg determines the mode of
segmentation; and its relative abundance affects profoundly the entire
embryonic history, and decides at what particular stage, and of what
size and form, the embryo shall hatch.

The loss of food yolk is another disturbing element, the full influence
of which is as yet imperfectly understood, but the possibility of which
must be always kept in mind. It is best known in the case of mammals,
where it has led to apparent, though very deceptive, simplification of
development; and it will probably not be until the embryology of the
large-yolked monotremes is at length described, that we shall fully
understand the formation of the germinal layers in the higher placental
mammals.

Amongst invertebrates we know but little as yet concerning the

[1] P. and F. Sarasin, *Ergebnisse naturwissenschaftlicher Forschungen auf
Ceylon.* Bd. ii. Heft iii. 1889.
[2] *Cf.* E. B. Wilson, 'The Development of Renilla.' *Phil. Trans.* 1883.
p. 755.

effects of loss of food yolk. It has been suggested that the extraordinary
nature of the segmentation of the egg of *Peripatus capensis*, made
known to us through Mr. Sedgwick's admirable researches, may be due
to loss of food yolk ; a suggestion which receives support from the
long duration of uterine development in this case.

Our knowledge is very imperfect as to the ease with which food
yolk may be acquired or lost ; but until our information is more
precise on this point, it seems unwise to lay much stress on suggested
pedigrees which involve great and frequent alternations in the amount
of food yolk present.

Of causes other than food yolk, or only indirectly connected with it,
which tend to falsify the ancestral history, many are now known, but
time will only permit me to notice the more important.   These are
distortion, whether in time or space ; sudden or violent metamorphosis ;
a series of modifications, due chiefly to mechanical causes, and which
may be spoken of as developmental conveniences ; the important
question of variability in development ; and finally the great problem
of degeneration.

Concerning distortions in time, all embryologists have noticed the
tendency to anticipation or precocious development of characters which
really belong to a later stage in the pedigree. The early attainment of
the cyclical form in the shell of *Orbitolites complanata* is a case in point ;
and Würtenberger has specially noticed this tendency in Ammonites.
Many early larvæ show it markedly, the explanation in this case being
that it is essential for them to hatch in a condition capable of
independent existence, *i.e.*, capable, at any rate, of obtaining and
digesting their own food.

Anachronisms, or actual reversal of the historical order of development of organs or parts, occur frequently.   Thus the joint surfaces of
bones acquire their characteristic curvatures before movement of one
part on another is effected, and before even the joint cavities are
formed.

Another good example is afforded by the development of the
mesenterial filaments in Alcyonarians.   Wilson has shown in the case
Renilla that in the development of an embryo from the egg the
six endodermal filaments appears first, and the two long ectodermal
filaments at a later period ; but that in the formation of a bud this

order of development is reversed, the first formed. He suggests, in explanation, that as filaments are the digestive organs, it is of primary importance to free embryo that they should be formed quickly. The long ectoderma filaments are chiefly concerned with maintaining currents of through the colony; in bud development they appear before the endodermal filaments, because they enable the bud during its early stages to draw nutrient matter from the body fluid of the parent while the endodermal filaments cannot come into use until the bud ha acquired both mouth and tentacles.

The completion of the ventricular septum in the heart of vertebrates before the auricular septum is a well-known anachronism and every embryologist could readily furnish many other cases.

A curious instance is afforded by the development of the teeth in mammals, if recent suggestions as to the origin of the milk dentition are confirmed, and the milk dentition prove to be a more recent acquisition than the permanent one.[1]

But the most important cases in reference to distortion in time concern the reproductive organs. If development were a strict and correct recapitulation of ancestral history, then each stage would possess reproductive organs in a mature condition. This is not the case, and it is clearly of the greatest importance that it should not be. It is true that the first commencement of the reproductive organs may occur at a very early larval stage, or even that the very first step in development may be a division of the egg into somatic and reproductive cells; and it is possible that, as maintained by Weismann, this condition is a primitive one. Still, even in these cases the reproductive organs merely commence their development at these early stages, and do not become functional until the animal is adult.

Exceptionally in certain animals, and as a normal occurrence in others, precocious maturation of the reproductive organs takes place, and a larval form becomes capable of sexual reproduction. These may lead to arrest of development, either at a late larval the Axolotl, or at successively earlier and earlier in the

[1] Cf. Oldfield, Thomas. '... the Homologies and Successions in the Teeth in the Docarxata, with an attempt to trace ... in the Mammalian teeth in general.' ...

gonophores of the Hydromedusae, until finally the extreme condition seen in Hydra is produced.

We do not know the causes that determine the period, whether late or early, at which the reproductive organs ripen, but the question is one of great interest and importance and deserves careful attention. The suggestion has been made that entire groups of animals, such as the Mesozoa, are merely larvae, arrested through such precocious acquiring of reproductive power, and it is conceivable that this may be the case. Mesozoa are a puzzling group in which the life history, though known with tolerable completeness, has as yet given us no reliable clue concerning their affinities to other animals, a tantalising distinction that is shared with them by Rotifers and Polyzoa.

Distortion of a curious kind it seen in cases of abrupt metamorphosis, where, as in the case of many Echinoderms, of Phoronis, and of the metabolic insects, the larva and the adult differ greatly in form, habits, mode of life, and very usually in the nature of their food and the mode of obtaining it : and the transition from one stage to the other is not a gradual but an abrupt one, at any rate so far as external characters are concerned.

Sudden changes of this kind, as from the free swimming Pluteus to the creeping Echinus, or from the sluggish leaf-eating caterpillar to the dainty butterfly, cannot possibly be recapitulatory, for even if small jumps are permissible in nature, there is no room for bounds forward of this magnitude. Cases of abrupt metamorphosis may always be viewed as due to secondary modifications, and rarely, if ever, have any significance beyond the particular group of animals concerned. For example, a Pluteus larva may be recognised as belonging to the group of Echinoidea before the adult urchin has commenced to be formed within it, and the Lepidopteran caterpillar is already an unmistakable insect. Hence, for the explanation of the metamorphoses in these cases it is useless to look outside the groups of Echinoidea and Insecta respectively.

Abrupt metamorphosis is always associated with great change in external form and appearance, and in mode of life, and very usually in nutrition. A gradual transition in such cases is inadmissible, because in the intermediate stages the animal would be adapted to neither the larva nor the adult condition ; a gradual conversion of the biting

mouth parts of the caterpillar to the sucking
would inevitably lead to starvation. The
retaining the external form and habits of one   an
unduly long period, so that the relations of the
surrounding environment remain unchanged, while
tions for the later stage are in progress. Cinderella and the princess
are equally possible entities, each being well adapted to her environment.
The exigencies of the situation do not permit, however, of a gradual
change from one to the other; and the transformation, at least as
regards external appearance, must be abrupt.

Kleinenberg has recently directed attention to cases in which the
larval and adult organs develop independently; the larval nervous
system, for instance, aborting completely and forming no part of that
of the adult. I am not sure that I fully understand Kleinenberg's
argument, but it seems very possible that such cases, which are
probably far more numerous than is yet admitted, may be due to what
may be termed the telescoping of ancestral stages one with another,
which takes place in actual development, and may accordingly be
grouped under the head of developmental convenience. Undue
prolongation of an early ancestral stage, as in cases of abrupt metamorphosis, must involve modification, especially in the muscular and
nervous systems; in such cases a telescoping of ancestral stages takes
place as we have seen, the adult being developed within the larva.
Such telescoping must distort the recapitulatory history, and the shape
of the larva and adult may differ widely, an independent origin of
organs, especially the muscular and nervous systems, may be acquired
secondarily.

The stage in the development of Squilla, in which the three
posterior maxillipedes disappear completely, to reappear at
in a totally different form, is not to be interpreted as meaning
adult maxillipedes are entirely new structures unconnected
with those of the larva. Neither is the annual shedding of
of deer to be regarded as the repetition of an ancestral
dition intercalated historically between successive stages
antlers. In both cases the explanation is afforded by
whether of the embryo or adult.

Many embryological modifications or distortions may

to mechanical causes, and may fairly be considered under the head of developmental conveniences

The amnion of higher vertebrates is a case in point, and is probably rightly explained as due in the first instance to sinking or depression of the embryo into the yolk, in order to avoid distortion through pressure against a hard unyielding eggshell. A similar device is employed, presumably for the same reason, in the early development of many insect embryos; and the depression of the Tænia head within the cyst is a phenomenon of very similar nature.

Restriction of the space within which development occurs often causes displacement or distortion of organs whose growth, restricted in its normal direction, takes place along the lines of least resistance. The telescoping of the limbs and other organs within the body of an insect larva is a simple case of such distortion; and a more complicated example, closely comparable in many ways to the invagination of the Tænia head, is afforded by the remarkable inversion of the germinal layers in Rodents, first described by Bischoff in the Guinea pig, and long believed to be peculiar to that animal, but subsequently and simultaneously discovered by three independent observers, Kupffer, Selenka, and Fraser, to occur in varying degrees in rats, mice, and in other rodents.

One of the most recent attempts to explain developmental peculiarities as due to mechanical causes is Mr. Dendy's suggestion with regard to the pseudogastrula stage in the development of the calcareous sponges. It is well known that while the larva is in the amphiblastula stage, and still imbedded in the tissues of the parent, the granular cells become invaginated within the ciliated cells, giving rise to the pseudogastrula stage. At a slightly later stage, when the larva becomes free, the invaginated granular cells become again everted, and the larva spherical in shape; while still later invagination occurs once more, the ciliated cells being this time invaginated within the granular cells. The significance of the pseudogastrula stage has hitherto been undetermined, but Mr. Dendy points out that the larva always occupies a definite position with reference to the parental tissues; that the ciliated half of the larva is covered by a soft and yielding wall, while the opposite half, composed of the granular cells, is covered by a layer stiffened with rigid spicules; and his observations on the growth of the larva lead him to think that the pseudogastrula stage is brought about occasionally to the value of the granular ... of spicules.

Lastly ... us with many ... to be wondered at that this should be the case. Some of ... fairly be spoken of as mere curiosities of development, ... clearly of greater moment. I do not propose to ... will merely mention two or three which I happen ... my head against and remember vividly.

The solid condition of the œsophagus, in Elasmobranch ... first noticed by Balfour, is a very curious point. The ... at first a well-developed lumen, like the rest of the alimentary ... but at an early period, stage K of Balfour's nomenclature, the ... the œsophagus overlying the heart, and immediately behind ... branchial region, becomes solid, and remains solid for a long time, the exact date of reappearance of the lumen not being yet ascertained.

Mr. Bles and myself have recently noticed a similar solidification of the œsophagus occurs in tadpoles of the common frog. In young free swimming tadpoles the œsophagus is perforate, but in tadpoles of about $7\frac{1}{2}$ mm. length it becomes solid and remains so until a length of about $10\frac{1}{2}$ mm. has been attained. The solidification occurs at a stage closely corresponding with that in which it first appears in the dogfish, and a curious point about it is that in the frog the œsophagus ... just before the mouth opening is formed, and remains solid for some little time after this important event.

This closing of the œsophagus clearly cannot be recapitulation, but the fact that it occurs at corresponding periods in the frog ... suggests that it may possibly, as Balfour hinted, 'turn out to have some unsuspected morphological bearing.'

Another developmental curiosity is the duplication of ... by growth downwards of tongues from their dorsal margins; ... cation which is described as occuring in Amphioxus and in ... but in no other animal; and the occurrence of which, in ... closely similar fashion, is one of the strongest arguments ... real affinity between these two forms. It is hardly ... a modification should have been acquired independentl ...

A much more litigious question is the ... canal of vertebrates, that curious tubular ...

central canal of the nervous system and the hinder end of the
alimentary canal that is conspicuously present in the embryos of lower
vertebrates, and retained in a more or less disguised condition in
the higher groups as well.

The neurenteric canal was discovered by that famous embryologist
Kowalevsky in Ascidians and Amphioxus. He drew special attention to
the occurrence of a stage in both Ascidians and Amphioxus in which the
larva is free swimming and in which the sole communication between
the alimentary cavity and the exterior is through the neurenteric canal
and the central canal of the nervous system; and suggested[1] that
animals may have existed or may still exist in which the nerve tube
fulfilled a non-nervous function, and possibly acted as part of the
alimentary canal; a suggestion that has recently been revived in a
somewhat extravagant form.

A passage of food particles into the alimentary cavity through the
neural tube has not yet been seen, and probably does not occur, as the
larva still possesses sufficient food yolk to carry it on in its development. It is therefore permissible to hold that the neurenteric canal
may be a mere embryological device, and devoid of any deep morphological significance.

The question of variation in development is one of very great
importance, and has perhaps not yet received the attention it deserves.
We are in some danger of assuming tacitly that the mode of development of allied animals will necessarily agree in all important respects
or even in details, and that if the development of one member of
a group be known, that of the others may be assumed to be similar.
The more recent progress of embryology is showing us showing us that
such inferences are not safe, and that in allied genera or species, or
even in different individuals of the same species, variations of
development may occur affecting important organs and at almost any
stage in their formation.

Great individual variations in the earliest processes of development,
i.e., the segmentation of the egg, have been described by different
writers.

[1] A. Kowalevsky, 'Weitere Studien über die Entwickelungs-Geschichte
des Amphioxus lanceolatus;' *Archic für mikroskopische Anatomie*, Bd. xiii.
1877, p. 201.

In Renilla, Wilson found an
segmentation of eggs from which apparently
produced. In some cases the egg divided into two in the
manner, in other cases it divided at once
which in different specimens were
or markedly unequal in size. Sometimes a preliminary
occurred without any further result, the egg returning to
shape, and pausing for a time before   the attempt
segment. Segmentation sometimes
telolecithal eggs, with the formation of four or
rest of the egg breaking up later, either
gressively, into segments about equal in size to those first formed
while lastly, in some instances segmentation was very irregular, following no apparent law.

It is noteworthy that the variability in the case of Renilla is
apparently confined to the earliest stages, for whatever the
segmentation, the embryos in their later stages were indistinguishable
from one another.

Similar modifications in the segmentation of the egg have been
described in the oyster by Brooks, in Anodon and other Mollusca, in
Hydra, and in Lumbricus, in which last Wilson has recently shown
that marked differences occur in the eggs even of t   individual
animal. In the different species of Peripatus there appear also to be
considerable variations in the details of segmentation.

In the early embryonic stages after the completion of segmentation
very considerable variation may occur in allied species or genera.
Among Cœlenterates, for instance, the mode of formation of the
hypoblast presents most perplexing modifications: it may arise as a
true gastrula invagination; as cells budded off from one pole of the
blastula into its cavity; as cells budded off from various
wall of the blastula; by delamination or actual division of
the blastula wall; or it may be present from the start as
of cells enclosed by the epiblast. It is in connection
variations that controversy has arisen as to the
development of the gastrula, a point to which I

Among the   Mollusca or Cœlomata the
tions in the position   conceivable

mesoblast in different and often in closely allied forms have given rise to ardent discussion, and have led to the proposal of theory after theory, each rejected in turn as only affording a partial explanation, and now culminating in Kleinenberg's protest against the use of the term mesoblast at all, at any rate in a sense implying any possibility of comparison with the primary layers, epiblast, and hypoblast, of Cœlenterata.

This is not the place to attempt to decide so difficult and technical a point, even were I capable of so doing, but we may well take warning from this extraordinary diversity of development, the full extent of which I believe we as yet realise most imperfectly, that in our attempts to reconstruct ancestral history from ontogenetic development we have taken in hand no light task. To reconstruct Latin from modern European languages would in comparison be but child's play.

Of the readiness with which special developmental characters are acquired by allied animals the brothers Sarasin[1] have given us evidence in the extraordinary modifications presented by the embryonic and larval respiratory organs of Amphibians.

Confining ourselves to those forms which do not lay their eggs in water, and in which consequently development takes place within the egg, we find that Ichthyophis and Salamandra have three pairs of specially modified external gills. Nototrema has two pairs; Alytes and Typhlonectes have only a single pair, which in the latter genus take the form of enormous leaf-like outgrowths from the sides of the neck. In Hylodes and Pipa there are no gills, the tail acting as the larval respiratory organ; in *Rana opisthodon*, according to Boulenger, larval respiration is effected by nine pairs of folds of the skin of the ventral surface of the body.

Most of these extraordinarily diversified organs are clearly secondarily acquired structures; it is possible that they all are, and that external gills, as was suggested by Balfour for Elasmobranchs, are to be regarded as embryonal respiratory organs acquired by the larvæ and of no ancestral value. The point, however, cannot be considered settled, for on this view the external gills of Elasmobranchs and Amphibians would be independently acquired and not homologous structures, a

[1] P. and F. Sarasin, *Ergebnisse naturwissenschaftlicher Forschungen auf Ceylon*, vol. ii. chap. i. pp. 24-38.

view contradicted by the close agreement in the two groups, as well as by the absence of any and internal gills in Amphibians.

It is well known that the frog and the newt differ in important points of their development. The two-layered of the epiblast in the frog is a marked point of difference, which involves further changes in the mode of formation of the nervous system and sense organs. The kidneys and their ducts differ considerably in their development in the two forms, as do also the bloodvessels.

Concerning the early development of the bloodvessels, there are considerable differences even between the allied species of frogs. In *Rana esculenta* M——— finds that there is at first in each branchial arch a single or aortic arch, running directly from the heart to the aorta; from the cardiac end of this aortic arch a vessel grows out into the gill as the afferent branchial vessel, the original aortic arch losing its connection with the heart, and becoming the efferent branchial vessel. Afferent and efferent branchial vessels become connected by capillaries in the gill, and the course of the circulation, so long as gill-breathing is maintained, is from the heart through the truncus arteriosus to the afferent branchial vessel, then through the gill capillaries to the efferent branchial vessel, and then on to the aorta. When the pulmonary circulation is thoroughly established the branchial circulation is cut off by the efferent vessel reacquiring its connection with the heart, when the blood naturally takes the direct passage along it to the aorta, and so escapes the gill capillaries.

In *Rana temporaria* the mode of development is very different: the afferent and efferent vessels arise in each arch independently and almost simultaneously: the afferent vessel soon acquires connection with the heart; but, unlike *R. esculenta*, the efferent vessel has no connection with the heart until the gills are about to atrophy.

In other words, the continuous aortic arch, from heart to aorta, is present in *R. esculenta* prior to the development of the gills: it becomes interrupted while the gills are in functional use, but is re-established when these begin to atrophy. In *R. temporaria*, on the other hand, there is no continuous aortic arch until the gills begin to atrophy.

The difference is an important one, for it is a matter of considerable morphological interest to determine whether the continuous aortic arch is primitive for vertebrates : *i.e.*, whether it existed prior to the development of gills. This point could be practically settled if we could decide which of the two frogs, *R. esculenta* and *R. temporaria*, has most correctly preserved its ancestral history in this respect.

About this there can be little doubt. The development of the vessels in the newts, a less modified group than the frogs, agrees with that of *R. esculenta*, and interesting confirmation is afforded by a single aberrant specimen of *R. temporaria*, in which Mr. Bles and myself found the vessels developing after the type of *R. esculenta*, *i.e.*, in which a complete aortic arch was present before the gills were formed.

We are, therefore, justified in concluding that as regards the development of the branchial bloodvessels, *R. esculenta* has retained a primitive ancestral character which is lost in *R. temporaria*, and it is interesting to note that were our knowledge of the development of amphibians confined to the common frog, the most likely form to be studied, we should, in all probability, have been led to wrong conclusions concerning the ancestral condition of the bloodvessels in a point of considerable importance.

A matter which at present is attracting much attention is the question of degeneration.

Natural selection, though consistent with and capable of leading to steady upward progress and improvement, by no means involves such progress as a necessary consequence. All it says is that those animals will, in each generation, have the best chance of survival which are most in harmony with their environment, and such animals will not necessarily be those which are ideally the best or most perfect.

If you go into a shop to purchase an umbrella the one you select is by no means necessarily that which most nearly approaches ideal perfection, but the one which best hits off the mean between your idea of what an umbrella should be and the amount of money you are prepared to give for it : the one, in fact, that is on the whole best suited to the circumstances of the case or the environment for the time being. It might well happen that you had a violent antipathy to a crooked handle, or else were determined to have a catch of a particular kind to secure the ribs, and this might lead to the selection, *i.e.*, the survival, of an article that in other and even in more important respects was manifestly inferior to the average.

also with animals : the survival of a   is   To animals living in possession of eyes is of no advantage, and forms not merely lose nothing thereby, but would actually they would escape the dangers that might arise delicate and complicated organ.   In extreme cases, leading a parasitic existence, the conditions of life may be such as render locomotor, digestive, sensory, and other organs entirely useless : in such cases those forms will be best in harmony with their surroundings which avoid the waste of energy resulting from the formation and maintenance of these organs.

Animals which have in this way fallen from the high estate of their forefathers, which have lost organs or systems which their progenitors possessed are commonly called degenerate.   The principle of degeneration, recognised by Darwin as a possible, and, under certain conditions, a necessary consequence of his theory of natural selection, has been since advocated strongly by Dohrn, and later by Lankester in an Evening Discourse delivered before the Association at the Sheffield Meeting in 1879.   Both Dohrn and Lankester suggested that degeneration occurred much more widely than was generally recognised.

In animals which are parasitic when adult, but free swimming in their early stages, as in the case of the Rhizocephala, whose life history was so admirably worked out by Fritz Muller, degeneration is clear enough : so also is it in the case of the solitary Ascidians, in which the larva is a free swimming animal with a notochord, an elongated tubular nervous system, and sense organs, while the adult is fixed, devoid of the swimming tail, with no notochord, and with a greatly reduced nervous system and aborted sense organs.

In such cases the animal, when adult, is, as regards the totality of its organisation, at a distinctly lower morphological   ly differentiated than it is when young, and during ment there is actual retrograde development of organs.

About such cases there is no doubt ; but we are asked to idea of degeneration much more widely.   It to   evidence   We are reminded of the tendency complete omission of ancestral stages of which we have quoted examples above; and it is suggested that if such larval stages were omitted in all the members of a group we should have no direct evidence of degeneration in a group that might really be in an extremely degenerate condition.

Supposing, for instance, the free larval stages of the solitary Ascidians were suppressed, say through the acquisition of food yolk, then it is urged that the degenerate condition of the group might easily escape detection. The supposition is by no means extravagant; food yolk varies greatly in amount in allied animals, and cases like Hylodes, or amongst Ascidians Pyrosoma, show how readily a mere increase in the amount of food yolk in the egg may lead to the omission of important ancestral stages.

The question then arises whether it is not possible, or even probable, that animals which now show no indication of degeneration in their development are in reality highly degenerate, and whether it is not legitimate to suppose such degeneration to have occurred in the case of animals whose affinities are obscure or difficult to determine.

It is more especially with regard to the lower vertebrates that this argument has been employed; and at the present day zoologists of authority, relying on it, do not hesitate to speak of such forms as Amphioxus and the Cyclostomes as degenerate animals, as wolves in sheep's clothing, animals whose simplicity is acquired and deceptive rather than real and ancestral.

I cannot but think that cases such as these should be regarded with some jealousy: there is at present a tendency to invoke degeneration rather freely as a talisman to extricate us from morphological difficulties; and an inclination to accept such suggestions, at any rate provisionally, without requiring satisfactory evidence in their support.

Degeneration of which there is direct embryological evidence stands on a very different footing from suspected degeneration, for which no direct evidence is forthcoming; and in the latter case the burden of proof undoubtedly rests with those who assume its existence.

The alleged instances among the lower vertebrates must be regarded particularly closely, because in their case the suggestion of degeneration is admittedly put forward as a means of escape from difficulties arising through theoretical views concerning the relation between vertebrates and invertebrates.

Amphioxus itself, so far as I can see, shows in   no
sign of degeneration, except possibly with regard to
diverticula, whose ultimate fate is not altogether clear.
to the earlier stages of development, concerning which, thanks to the
patient investigations of Kowalevsky and Hatschek, our knowledge is
precise, there is no animal known to us in which the sequence of events
is simpler or more straightforward. Its various organs and systems
are formed in what is recognised as a primitive manner ; and the
development of each is a steady upward progress towards the adult
condition. Food yolk, the great cause of distortion in development, is
almost absent, and there is not the slightest indication of the former
possession of a larger quantity. Concerning the later stages our
knowledge is incomplete, but so much as has been ascertained gives no
support to the suggestion of general degeneration.

Our knowledge of the conditions leading to degeneration is undoubtedly incomplete, but it must be noticed that the conditions
usually associated with degeneration do not occur. Amphioxus is
not parasitic, is not attached when adult, and shows no evidence of
having formerly possessed food yolk in quantity sufficient to have led
to the omission of important ancestral stages. Its small size as compared with other vertebrates is one of the very few points that can be
referred to as possibly indicating degeneration, and will be considered
more fully at a later point in my address.

A consideration of much less importance, but deserving of mention,
is that in its mode of life Amphioxus not merely differs as already
noticed from those groups of animals which we know to be degenerate,
but agrees with some, at any rate, of those which there is
regard as primitive or persistent types. Amphioxus,
Lingula, Dentalium, and Limulus, is marine, and occurs in
water, usually with a sandy bottom, and, like the three smaller of these
genera, it lives habitually buried almost completely in   into
which it burrows with great rapidity.

I do not wish to speak dogmatically. I merely wish
against a too ready assumption of degeneration ; and to
far as I can see, Amphioxus has not yet, either in its
its structure, or in its habits, been shown to present
suggest, still less that prove, the occurrence in it
tensive degeneration.

In a sense, all the higher animals are degenerate ; that is, they can be shown to possess certain organs in a less highly developed condition than their ancestors, or even in a rudimentary state.

Thus a crab as compared with a lobster is degenerate in the matter of its tail, a horse as compared with Hipparion in regard to its outer toes ; but it is neither customary nor advisable to speak of a crab as a degenerate animal compared to a lobster ; to do so would be misleading. An animal should only be spoken of as degenerate when the retrograde development is well marked, and has affected not one or two organs only, but the totality of its organisation.

It is impossible to draw a sharp line in such cases, and to limit precisely the use of the term degeneration. It must be borne in mind that no animal is at the top of the tree in all respects. Man himself is primitive as regards the number of his toes, and degenerate in respect to his ear muscles ; and between two animals even of the same group it may be impossible to decide which of the two is to be called the higher and which the lower form.

Thus, to compare an oyster with a mussel. The oyster is more primitive than the mussel as regards the position of the ventricle of the heart and its relations to the alimentary canal ; but is more modified in having but a single adductor muscle ; and almost certainly degenerate in being devoid of a foot.

Care must also be taken to avoid speaking of an animal as degenerate in regard to a particular organ merely because that organ is less fully developed than in allied animals. An organ is not degenerate unless its present possessor has it in a less perfect condition than its ancestors had.

A man is not degenerate in the matter of the length of his neck as compared with a giraffe, nor as compared with an elephant in respect of the size of his front teeth, for neither elephant nor giraffe enters into the pedigree of man. A man, is, however, degenerate, whoever his ancestors may have been, in regard to his ear muscles ; for he possesses these in a rudimentary and functionless condition, which can only be explained by descent from some better equipped progenitor.

Closely connected with the question of degeneration is that of the size of animals, and its bearing on their structure and development ; a problem noticed by many writers, but which has perhaps not yet received the attention it merits.

If we are right in interpreting the eggs of Metazoa as representing the unicellular or protozoan stage in their ancestry, then the small size of the egg may be viewed as recapitulatory.

But the gradual increase in size of the embryo, and its growth up to the adult condition, can only be regarded as representing in a most general way, if at all, the actual or even the relative sizes of the intermediate ancestral stages of the pedigree.

It is quite true that animals belonging to the lower groups are, as a general rule, of smaller size than those of higher grade; and also that the giants are met with among the highest members of each division. Cephalopoda are the highest molluscs, and the largest cephalopods greatly exceed in size any other members of the group; decapods are at once the highest and the largest crustaceans; and whales, the hugest animals that exist, or, so far as we know, that ever have existed, belong to the highest group of all, the mammalia. It would be easy to quote exceptions, but the general rule obtains admittedly.

However, although there may be, and probably is, a general parallelism between the increase in size from the egg to the adult, and the historical increase in size during the passage from lower to higher forms; yet no one could maintain that the sizes of embryos represent at all correctly those of the ancestors; that, for instance, the earliest birds were animals the size of a chick embryo at a time when avian characters first declare themselves, or that the ancestral series in all cases presented a steady progression in respect of actual magnitude.

In the lower animals, *e.g.*, in Orbitolites, the actual size of the several ancestral stages is probably correctly recapitulated during the growth of the adult; and it is very possible that it is so also in such forms as the solitary sponges. In higher animals, except in the early stages of those forms which are practically devoid of food yolk, and which hatch as pelagic larvæ, this certainly does not obtain.

This is clear enough, but is worth pointing out, for if, as most certainly is the case, the embryos of animals are actually smaller than the ancestral forms they represent, it is possible that the smallness of the embryo may have had some influence on its organisation, and be responsible for some of the modifications in the ancestral history, and more especially for the disappearance of ancestral organs in free swimming larvæ.

c

In adult animals the relation between size and structure has been very clearly pointed out by Herbert Spencer. Increased size involves by itself increased complexity of structure; the determining consideration being that while the surface area of the body increases as the squares of the linear dimensions, the mass of the body increases as their cubes.

If, for example, we imagine two animals of similar shape and proportions, but of different size; for the sake of simplicity, we may suppose them to be spherical, and that the diameter of one is twice that of the other; then the larger one will have four times the extent of surface of the smaller, but eight times its mass or bulk: and it is quite possible that while the extent of surface, or skin, in the smaller animal might suffice for the necessary respiratory and excretory interchanges, it would be altogether insufficient in the larger animal, in which increased extent of surface must be provided by foldings of the skin, as in the form of gills.

To take an actual instance; Limapontia is a minute nudibranchiate, or sea-slug, about the sixth of an inch in length; it has a smooth body, totally devoid of respiratory processes, while forms allied to it, but of larger size, have their extent of surface increased by branching processes, which often take the form of specialised gills.

This is a peculiarly instructive case, because Limapontia in its early developmental stages possesses a large spirally-coiled shell, and shows other evidence of descent from forms with specialised breathing organs. We are certainly right in associating the absence of respiratory organs in the adult with the small size of the animal; and comparison with allied forms suggests very strongly that there has been in its pedigree an actual reduction of size, which has led to the degeneration of the respiratory organs.

This is an important conclusion: it is a well-known fact that the smaller members of a group are, as a rule, more simply organised than the larger members, especially with regard to their respiratory and circulatory systems; but if we are right in concluding that reduction in size may be an actual cause of simplification or degeneration in structure, then we must be on our guard against assuming hastily that these smaller and simpler animals are necessarily primitive in regard to the groups to which they belong. It is possible, for instance, that the simplification or ——————— of respiratory organs seen in Pauropus, in the Thysanura, and ———————— Tracheata, may be a secondary character, acquired through reduction of size.

An interesting illustration of the law discussed above is afforded by the brains of mammals ; it has been noticed by many anatomists that the extent of convolution, or folding of the surface of the cerebral hemispheres in mammals, is related not to the degree of intelligence of the animal, but to its actual size, a beaver having an almost smooth brain and a cow a highly complicated one. Jelgersma, and independently of him, Professor Fitzgerald,[1] have explained this as due to the necessity of preserving the due proportion between the outer layer of grey matter or cortex, which is approximately uniform in thickness, and the central mass of white matter. But for the foldings of the surface the proportion of white matter to grey matter would be far higher in a large than in a small brain.

It must not be forgotten, on the other hand, that many zoologists hold the view, in favour of which the evidence is steadily increasing, that the primitive or ancestral members of each group were of small size. Thus Fürbringer remarks with regard to birds, that on the whole small birds show more primitive and simpler conditions of structure than the larger members of the same group. He expresses the opinion that the first birds were probably smaller than Archaeopteryx, and notes that reptiles and mammals also show in their earlier and smaller types more primitive features than do their larger descendants. Finally, Fürbringer concludes that 'it is therefore the study of the smaller members within given groups of animals which promises the best results as to their phylogeny.'

Again, one of the most striking points with regard to the pedigree of the horse, as agreed on by palæontologists, is the progressive reduction in size which we meet with as we pass backwards in time from stage to stage. The Pliocene Hipparion was smaller than the existing horse, in fact about the size of a donkey ; the Miocene Mesohippus about equalled a sheep ; while Eohippus, from the Lower Eocene deposits, was no larger than a fox. Not only is there good reason for holding that, as a rule, larger animals are descended from

[1] Cf. *Nature*, June 5, 1890, p. 125.

ancestors of smaller size, but there is also much evidence to show that
increase in size beyond certain limits is disadvantageous, and may lead
to destruction rather than to survival. It has happened more than
once in the history of the world, and in more than one group of
animals, that gigantic stature has been attained immediately before
extinction of the group, a final and tremendous effort to secure
survival, but a despairing and unsuccessful one. The Ichthyosauri,
Plesiosauri, and other extinct reptilian groups, the Moas, and the huge
extinct Edentates, are well-known examples, to which before long will
be added the elephants and the whales, and, it may be, ironclads
as well.

The whole question of the influence of size is of the greatest
possible interest and importance, and it is greatly to be hoped that it
will not be permitted to remain in its present uncertain and unsatisfactory condition.

It may be suggested that Amphioxus is an animal which has undergone reduction in size, and that its structural simplicity may, like that
of Limapontia, be due, in part at least, to this reduction. Such
evidence as we have tells against this suggestion ; the first system to
undergo degeneration in consequence of a reduction in size is the
respiratory, and the respiratory organs of Amphioxus, though very
simple, are also, for a vertebrate, unusually extensive.

We have now considered the more important of the influences
which are recognised as affecting developmental history in such a way
as to render the recapitulation of ancestral stages less complete than
it might otherwise be, which tend to prevent ontogeny from correctly
repeating the phylogenetic history. It may at this point reasonably
be asked whether there is any way of distinguishing the palingenetic
history from the later cenogenetic modifications grafted on to it ; any
test by which we can determine whether a given larval character is or
is not ancestral.

Most assuredly there is no one rule, no single test, that will apply in
all cases ; but there are certain considerations which will help us, and
which should be kept in view.

A character that is of general occurrence among the members of a
group, both high and low, may reasonably be regarded as having
strong claims to ancestral rank ; claims that are greatly strengthened

if it occurs at corresponding developmental periods in all cases; and
still more if it occurs equally in forms that hatch early as free larvæ,
and in forms with large eggs, which develop directly into the adult.
As examples of such characters may be cited the mode of formation
and relations of the notochord, and of the gill clefts of vertebrates,
which satisfy all the conditions mentioned.

Characters that are transitory in certain groups, but retained
throughout life in allied groups, may, with tolerable certainty, be
regarded as ancestral for the former; for instance, the symmetrical
position of the eyes in young flat fish, the spiral shell of the young
limpet, the superficial positions of the madreporite in Elasipodous
Holothurians, or the suckerless condition of the ambulacral feet in
many Echinoderms.

A more important consideration is that if the developmental changes
are to be interpreted as a correct record of ancestral history, then the
several stages must be possible ones, the history must be one that
could actually have occurred, i.e., the several steps of the history as
reconstructed must form a series, all the stages of which are practicable
ones.

Natural selection explains the actual structure of a complex organ
as having been acquired by the preservation of a series of stages,
each a distinct, if slight, advance on the stage immediately preceding
it, an advance so distinct as to confer on its possessor an appreciable
advantage in the struggle for existence. It is not enough that the
ultimate stage should be more advantageous than the initial or earlier
condition, but each intermediate stage must also be a distinct advance.
If then the development of an organ is strictly recapitulatory, it
should present to us a series of stages, each of which is not merely
functional, but a distinct advance on the stage immediately preceding
it. Intermediate stages, e.g., the solid œsophagus of the tadpole, which
which are not and could not be functional, can form no part of an
ancestral series; a consideration well expressed by Sedgwick[1] thus:
'Any phylogenetic hypothesis which presents difficulties from a
physiological standpoint must be regarded as very provisional indeed.'

[1] Sedgwick, 'On the Early Development of the Anterior Part of the
Wolffian Duct and Body in the Chick,' Quarterly Journal of Microscopical
Science, vol. xxi., 1881, p. 456.

A good example of an embryological series fulfilling these conditions is afforded by the development of the eye in the higher Cephalopoda. The earliest stage consists in the depression of a slightly modified patch of skin ; round the edge of the patch the epidermis becomes raised up as a rim ; this gradually grows inwards from all sides, so that the depressed patch now forms a pit, communicating with the exterior through a small hole or mouth. By further growth the mouth of the pit becomes still more narrowed, and ultimately completely closed, so that the pit becomes converted into a closed sac or vesicle ; at the point at which final closure occurs formation of cuticle takes place, which projects as a small transparent drop in the cavity of the sac ; by formation of concentric layers of cuticle this drop becomes enlarged into the spherical transparent lens of the eye, and the development is completed by histological changes in the inner wall of the vesicle, which convert it into the retina, and by the formation of folds of skin around the eye, which become the iris and the eyelids respectively.

Each stage in this developmental history is a distinct advance, physiologically, on the preceding stage, and, furthermore, each stage is retained at the present day as the permanent condition of the eye in some member of the group Mollusca.

The earliest stage, in which the eye is merely a slightly depressed and slightly modified patch of skin, represents the simplest condition of the Molluscan eye, and is retained throughout life in Solen. The stage in which the eye is a pit, with widely open mouth, is retained in the limpet ; it is a distinct advance on the former, as through the greater depression the sensory cells are less exposed to accidental injury.

The narrowing of the mouth of the pit in the next stage is a simple change, but a very important step forwards. Up to this point the eye has served to distinguish light from darkness, but the formation of an image has been impossible. Now, owing to the smallness of the aperture, and the pigmentation of the walls of the pit which accompanies the change, light from any one part of an object can only fall on one particular part of the inner wall of the pit or retina, and so an image, though a dim one, is formed. This type of eye is permanently retained in the Nautilus.

The closing of the mouth of the pit by a transparent membrane will not affect the optical properties of the eye, and will be a gain, as it will prevent the entrance of foreign bodies into the cavity of the cup.

The formation of the lens by deposit of cuticle is the next step. The gain here is increased distinctness and increased brightness of the image, for the lens will focus the rays of light more sharply on the retina, and will allow a greater quantity of light, a larger pencil of rays from each part of the object, to reach the corresponding part of the retina. The eye is now in the condition in which it remains throughout life in the snail and other gastropods. Finally the formation of the folds of skin known as iris and eyelids provides for the better protection of the eye, and is a clear advance on the somewhat clumsy method of withdrawal seen in the snail.

The development of the vertebrate liver is another good but simpler example. The most primitive form of the liver is that of Amphioxus, in which it is present as a simple saccular diverticulum of the intestinal canal, with its wall consisting of a single layer of cells, and with bloodvessels on its outer surface. The earliest stage in the formation of the liver in higher vertebrates, the frog for instance, is practically identical with this. In the frog the next stage consists in folding of the wall of the sac, which increases the efficiency of the organ by increasing the extent of surface in contact with the bloodvessels. The adult condition is attained simply by a continuance of this process; the foldings of the wall becoming more and more complicated, but the essential structure remaining the same—a single layer of epithelial cells in contact on one side with bloodvessels, and bounding on the other directly or indirectly the cavity of the alimentary canal.

It is not always possible to point out the particular advantage gained at each step even when a complete developmental series is known to us, but in such cases as, for instance, in Orbitolites, our difficulties arise chiefly from ignorance of the particular conditions that confer advantage in the struggle for existence in the case of the forms we are dealing with.

The early larval stages in the development of animals, and more especially those that are marine and pelagic in habit, have naturally

attracted much attention, since in the absence, probably inevitable, of
satisfactory palæontological evidence, they afford us the sole available
clue to the determination of the mutual relations of the large groups
of animals, or of the points at which these diverged from one another.

In attempting to interpret these early ontogenetic stages as actual
ancestral forms, beyond which development at one time did not
proceed, we must keep clearly in view the various disturbing causes
which tend to falsify the ancestral record; such as the influence of
food yolk, or of habitat, and the tendency of diminution in size to give
rise to simplification of structure, a point of importance if it be granted
that these free larvæ are of smaller size than the ancestral forms to
which they correspond.

If, on the other hand, in spite of these powerful modifying causes,
we do find a particular larval form occurring widely and in groups
not very closely akin, then we certainly are justified in attaching
great importance to it, and in regarding it as having strong claims to
be accepted as ancestral for these groups.

Concerning these larval forms, and their possible ancestral
significance, our knowledge has made no great advance since the
publication of Balfour's memorable chapter on this subject; and I
propose merely to allude briefly to a few of the more striking
instances.

The earliest, the most widely spread, and the most famous of larval
forms is the gastrula, which occurs in a simple or in a modified form
in some members of each of the large animal groups. It is generally
admitted that its significance is the same in all cases, and the evidence
is very strong in favour of regarding it as a stage ancestral for all
Metazoa. The difficulty arising from its varying mode of development in different forms is, however, still unsolved and embryologists
are not yet agreed whether the invaginate or delaminate form is the
more primitive. In favour of the former is its much wider occurrence;
in favour of the latter the fact that it is easy to picture a series of
stages leading gradually from a unicellular protozoon to a blastula, a
diblastula, and ultimately a gastrula, each stage being a distinct
advance, both morphological and physiological, on the preceding
stage; while in the case of the invaginate gastrula it is not easy to
imagine any advantage resulting from a flattening or slight pitting in

of one part of the surface, sufficient to lead to its preservation and further development.

Of larval forms later than the gastrula, the most important by far is the Pilidium larva, from which it is possible, as Balfour has shown, that the slightly later Echinoderm larva, as well as the widely spread Trochosphere larva, may both be derived. Balfour concludes that the larva forms of all Cœlomata, excluding the crustacea and vertebrates, may be derived from one common type, which is most nearly represented now by the Pilidium larva, and which 'was an organism something like a Medusa, with a radial symmetry.' The tendency of recent phylogenetic speculations is to accept this in full, and to regard as the ancestor of Turbellarians and of all higher forms, a jelly-fish or ctenophoran, which in place of swimming freely has taken to crawling on the sea bottom.

Of the two groups excluded above, the crustacea and the vertebrata, the interest of the former centres in the much discussed problem of the significance of the Nauplius larva. There is now a fairly general agreement that the primitive crustacea were types akin to the phyllopods, i.e., forms with elongated and many-segmented bodies, and a large number of pair of similar appendages. If this is correct, then the explanation of the Nauplius stage must be afforded by the phyllopods themselves, and it is no use looking beyond this group for it. A Nauplius larva occurs in other crustacea merely because they have inherited from their phyllopod ancestors the tendency to develop such a stage, and it is quite legitimate to hold that higher crustaceans are descended from phyllopods, and that the Nauplius represents in more or less modified form an earlier ancestor of the phyllopods themselves.

As to the Nauplius itself the first thing to note is that though an early larval form it cannot be a very primitive form, for it is already an unmistakable crustacean ; the absence of cilia, the formation of a cuticular investment, the presence of jointed schizopodous limbs, together with other anatomical characters, proving this point conclusively. It follows, therefore, either that the earlier and more primitive stages are entirely omitted in the development of crustacea, or else that the Nauplius represents such an early ancestral stage with crustacean characters, which properly belong to a later stage, thrown back upon it and precociously developed.

The latter explanation is the one usually adopted; but before the question can be finally decided more accurate observations than we at present possess are needed concerning the stages intermediate between the egg and the Nauplius.

The absence of a heart in the Nauplius may reasonably be associated with the small size of the larva.

Concerning the larval forms of vertebrates, it is only in Amphioxus and the Ascidians that the earliest larval stages are free-living, independent animals. In both groups the most characteristic larval stage is that in which a notochord is present, and a neural tube, open in front, and communicating behind through a neurenteric canal with the digestive cavity, which has no other opening to the exterior. This is a very early stage, both in Amphioxus and Ascidians; but, so far as we know, it cannot be compared with any invertebrate larva. It is customary, in discussions on the affinities of vertebrates, to absolutely ignore the vertebrate larval forms, and to assume that their peculiarities are due to precocious development of vertebrate characteristics. It may turn out that this view of the matter is correct; but it has certainly not yet been proved to be so, and the development of both Amphioxus and Ascidians is so direct and straightforward that evidence of some kind may reasonably be required before accepting the doctrine that this development is entirely deceptive with regard to the ancestry of vertebrates.

Zoologists have not quite made up their minds what to do with Amphioxus: apparently the most guileless of creatures, many view it with the utmost suspicion, and not merely refuse to accept its mute protestations of innocence, but regard and speak of it as the most artful of deceivers. Few questions at the present day are in greater need of authoritative settlement.

That ontogeny really is a repetition of phylogeny must, I think, be admitted, in spite of the numerous and various ways in which the ancestral history may be distorted during actual development.

Before leaving the subject, it is worth while inquiring whether any explanation can be found of recapitulation. A complete answer can certainly not be given at present, but a partial one may, perhaps, be obtained.

Darwin himself suggested that the clue might be found in the consideration that at whatever age a variation first appears in the parent, it tends to reappear at a corresponding age in the offspring ; but this must be regarded rather as a statement of the fundamental fact of embryology than as an explanation of it.

It is probably safe to assume that animals would not recapitulate unless they were compelled to do so : that there must be some constraining influences at work, forcing them to repeat more or less closely the ancestral stages. It is impossible for instance to conceive what advantage it can be to a reptilian or mammalian embryo to develop gill clefts which are never used, and which disappear at a slightly later stage ; or how it can benefit a whale, that in its embryonic condition it should possess teeth which never cut the gum, and which are lost before birth.

Moreover, the history of development in different animals or groups of animals, offers to us, as we have seen, a series of ingenious, determined, varied, but more or less unsuccessful efforts to escape from the necessity of recapitulating, and to substitute for the ancestral process a more direct method.

A further consideration of importance is that recapitulation is not seen in all forms of development, but only in sexual development ; or, at least, only in development from the egg. In the several forms of asexual development, of which budding is the most frequent and most familiar, there is no repetition of ancestral phases ; neither is there in cases of regeneration of lost parts, such as the tentacle of a snail, the arm of a starfish, or the tail of a lizard ; in such regeneration it is not a larval tentacle, or arm, or tail, that is produced, but an adult one.

The most striking point about the development of the higher animals is that they all alike commence as eggs. Looking more closely at the egg and the conditions of its development, two facts impress us as of special importance : first, the egg is a single cell, and therefore represents morphologically the Protozoan, or earliest ancestral phase ; secondly, the egg, before it can develop, must be fertilised by a spermatozoon, just as the stimulus of fertilisation by the pollen grain is necessary before the ovum of a plant will commence to develop into the plant embryo.

The advantage of cross-fertilisation in increasing the vigour of the offspring is well known, and in plants devices of the most varied and even extraordinary kind are adopted to ensure that such cross-fertilisation occurs. The essence of the act of cross-fertilisation, which is already established among Protozoa, consists in combination of the nuclei of two cells, male and female, derived from different individuals. The nature of the process is of such a kind that two individual cells are alone concerned in it; and it may, I think, be reasonably argued that the reason why animals commence their existence as eggs, i.e. as single cells, is because it is in this way only that the advantage of cross-fertilisation can be secured, an advantage admittedly of the greatest importance, and to secure which natural selection would operate powerfully.

The occurrence of parthenogenesis, either occasionally or normally, in certain groups is not, I think, a serious objection to this view. There are very strong reasons for holding that parthenogenetic development is a modified form, derived from the sexual method. Moreover, the view advanced above does not require that cross-fertilisation should be essential to individual development, but merely that it should be in the highest degree advantageous to the species, and hence leaves room for the occurrence, exceptionally, of parthenogenetic development.

If it be objected that this is laying too much stress on sexual reproduction, and on the advantage of cross-fertilisation, then it may be pointed out in reply that sexual reproduction is the characteristic and essential mode of multiplication among Metazoa : that it occurs in all Metazoa, and that when asexual reproduction, as by budding, &c., occurs, this merely alternates with the sexual process which, sooner or later, becomes essential.

If the fundamental importance of sexual reproduction to the welfare of the species be granted, and if it be further admitted that Metazoa are descended from Protozoa, then we see that there is really a constraining force of a most powerful nature compelling every animal to commence its life history in the unicellular condition, the only condition in which the advantage of cross-fertilisation can be obtained ; i.e., constraining every animal to begin its development at its earliest ancestral stage, at the very bottom of its genealogical tree.

On this view the actual development being denied is strictly reached at both ends: it must commence as an egg, and it must end in the likeness of the parent. The problem of recapitulation becomes thereby greatly narrowed; all that remains being to explain why the intermediate stages in the actual development should repeat the intermediate stages of the ancestral history.

Although narrowed in this way, the problem still remains one of extreme difficulty.

It is a consequence of the Theory of Natural Selection that identity of structure involves community of descent: a given result can only be arrived at through a given sequence of events: the same morphological goal cannot be reached by two independent paths. A negro and a white man have had common ancestors in the past; and it is through the long-continued action of selection and environment that the two types have been gradually evolved. You cannot turn a white man into a negro merely by sending him to live in Africa: to create a negro the whole ancestral history would have to be repeated; and it may be that it is for the same reason that the embryo must repeat or recapitulate its ancestral history in order to reach the adult goal.

I am not sure that we can at present get much further; but the above considerations give opportunity for brief notice of what is perhaps the most noteworthy of recent embryological papers, Kleinenberg's remarkable monograph on Lopadorhynchus.

Kleinenberg directs special attention to what is known to evolutionists as the difficulty with regard to the origin of new organs, which is to the effect that although natural selection is competent to account for any amount of modification in an organ after it has attained a certain size, and become of functional importance, yet that it cannot account for the earliest stages in the formation of an organ before it has become large enough or sufficiently developed to be of real use. The difficulty is a serious one; it is carefully considered by Mr. Darwin, and met completely in certain cases; but as Kleinenberg correctly states, no general explanation has been offered with regard to such instances.

As such general explanation Kleinberg proposes his Theory of the development of organs by substitution. He points out that any modification of an organ or tissue must involve modification at least in functional activity, of other organs. He then continues by urging that one organ may replace or be substituted for another, the replacing organ being in no way derived morphologically from the replaced or preceding organ, but having a genetic relation to it of this kind :— that it can only arise in an organism so constituted, and is dependent on the prior existence of the replaced organ, which supplies the necessary stimulus for its formation.

As an example he takes the axial skeleton of vertebrates. The notochord, formed by change of function from the wall of the digestive canal, is the sole skeleton of the lowest vertebrates, and the earliest developmental phase in all the higher forms. The notochord gives rise directly to no other organ, but is gradually replaced by other and unlike structures by substitution. The notochord is an intermediate organ, and the cartilaginous skeleton which replaces it is only intelligible through the previous existence of the notochord ; while, in its turn, the cartilaginous skeleton gives way, being replaced, through substitution, by the bony skeleton.

The successive phases in the evolution of weapons might be quoted as an illustration of Kleinenberg's theory. The bow and arrow is a better weapon than a stick or stone ; it is used for the same purpose, and the importance or need for a better weapon led to the replacement of the sling by the bow ; the bow does not arise by further development or increasing perfection of the sling : it is an entirely new weapon, towards the formation of which the older and more primitive weapons have acted as a stimulus, and which has replaced these latter by substitution, while the substitution at a later date of firearms for the bow and arrow is merely a further instance of the same principle.

It is too early yet to realise the full significance of Kleinenberg's most suggestive theory ; but if it be really true that each historic stage in the evolution of an organ is necessary as a stimulus to the development of the next succeeding stage, then it becomes clear why animals are constrained to recapitulate. Kleinenberg suggests further that the extraordinary persistence in embryonic life of organs which are rudimentary and functionless in the adult may also be explained by his theory, the presence of such organs in the embryo being indispensable as a stimulus to the development of the permanent structures of the adult.

It would be easy to point out difficulties in the way of the theory. The omission of historic stages in the actual ontogenetic development, of which almost all groups of animals supply striking examples, is one of the most serious; for if these stages are necessary stimuli for the succeeding stages, then their omission is unexplained; while if such stimuli are not necessary, the theory would appear to need revision.

Such objections may, however, prove to be less serious than they appear at first sight; and in any case Kleinenberg's theory may be welcomed as an important and original contribution, which deserves—indeed demands—the fullest and most careful consideration from all morphologists, and which acquires special interest from the explanation which it offers of recapitulation as a mechanical process, through which alone is it possible for an embryo to attain the adult structure.

That recapitulation does actually occur, that the several stages in the development of an animal are inseparably linked with and determined by its ancestral history, must be accepted. 'To take any other view is to admit that the structure of animals and the history of their development form a mere snare to entrap our judgment.'

Embryology, however, is not to be regarded as a master-key that is to open the gates of knowledge and remove all obstacles from our path without further trouble on our part; it is rather to be viewed and treated as a delicate and complicated instrument, the proper handling of which requires the utmost nicety of balance and adjustment, and which, unless employed with the greatest skill and judgment, may yield false instead of true results.

Embryology is indeed a most powerful and efficient aid, but it will not, and cannot, provide us with an immediate or complete answer to the great riddle of life. Complications, distortions, innumerable and bewildering, confront us at every step, and the progress of knowledge has so far served rather to increase the number and magnitude of these pitfalls than to teach us how to avoid them.

Still, there is no cause for despair—far from it; if our difficulties are increasing, so also are our means of grappling with them; if the goal appears harder to reach than we thought for, on the other hand its position is far better defined, and the means of approach, the lines of attack, are more clearly recognised.

One thing above all is apparent, that embryologists must not work single-handed, and must not be satisfied with an acquaintance, however exact, with animals from the side of development only ; for embryos have this in common with maps, that too close and too exclusive a study of them is apt to disturb a man's reasoning power.

Embryology is a means, not an end.   Our ambition is to explain in what manner and by what stages the present structure of animals has been attained.   Towards this embryology affords most potent aid ; but the eloquent protest of the great anatomist of Heidelberg must be laid to heart, and it must not be forgotten that it is through comparative anatomy that its power to help is derived.

What would it profit us, as Gegenbaur justly asks, to know that the higher vertebrates when embryos have slits in their throats, unless through comparative anatomy we were acquainted with forms now existing in which these slits are structures essential to existence? Anatomy defines the goal, tells us of the things that have to be explained ; embryology offers a means, otherwise denied to us, of attaining it.

Comparative anatomy and palæontology must be studied most earnestly by those who would turn the lessons of embryology to best account, and it must never be forgotten that it is to men like Johannes Müller, Stannius, Cuvier, and John Hunter, the men to whom our exact knowledge of comparative anatomy is due, that we owe also the possibility of a science of embryology.

PLATE I.

By OSWALD H. LATTER, M.A., *Formerly Berkeley Fellow of Owens College, Manchester, 1888, Late Tutor of Keble College, Assistant Master at Charterhouse.*

[*From the* PROCEEDINGS OF THE ZOOLOGICAL SOCIETY OF LONDON, January 20, 1891.]

I. *The Passage of the Ova from the Ovary to the External Gill-plate.*

In 1830 von Baer gave in Meckel's 'Archiv,' 1830, pp. 313–352, an account of this process, which has, so far as I can ascertain, been tacitly accepted by all later writers on the subject. My own observations have led me to somewhat different conclusions. Von Baer's account is briefly as follows :—The ova pass along the inner branchial passage, being prevented from falling into the internal gill-space by the labour contractions of the foot ; thence they pass into the cloaca, into which the outer branchial passage also opens. All the muscles of the body are in a state of contraction during the passage of the ova, and furthermore the cloaca is small. In consequence of the muscular contraction the shell is closed and the ova accumulate in the cloaca, a few perhaps being emitted into the water before the closure is complete. The only direction therefore along which the pressure of ova can be relieved is forwards along the outer branchial passage and thus to the external gill-space. It is to be noticed that von Baer does not state that he has *observed* these phenomena, but merely draws his conclusions from the anatomical relations of the various organs.

I have myself observed the passage of ova as far as the cloaca. The genital aperture, as is well known, is situated ventral of and somewhat posterior to the external aperture of the nephridium ; it is slightly anterior to the commencement of the free detached dorsal border of the inner lamella of the internal gill-plate. The ova may be seen through the thin epithelial covering on the dorsal margin of

D

the foot, passing along the oviduct to the genital aperture. After escaping through this pore they are conveyed backwards along the external surface of the nephridium. This surface is densely covered with cilia borne upon tall columnar cells, with a large oval nucleus lying in their lower portion and resting on a definite basement membrane. In the middle line of the nephridial surface the cilia are longer and drive the ova straight backwards; towards the ventral and dorsal sides of the nephridial surface the cilia are shorter and drive the ova obliquely backwards and towards the line of the longer cilia, so that the latter tend to keep the ova in the middle line where the ciliary currents are strongest. The arrows (Plate I. fig. 6) show the direction of the currents. The total effect of the cilia is therefore to drive the ova straight backwards along the middle line of the nephridial surface. In the course of about 50 seconds an ovum is thus swept back to the slit between the retractor pedis muscle and the point of fusion of the internal gill-plates. Through this slit the ova pass, meet the stream of ova from the other side of the body, and so reach the exhalant branchial current and the cloaca. The process goes on for several days (10 or 11) in each individual. This being the case, according to von Baer's theory the shell must remain closed during the whole of this period, or, in other words, respiration be suspended for nearly a fortnight. This appears to me incompatible with the continued life of the individual.

In order that the ova may reach their final resting-place there must be some reversal of the respiratory currents. I was unable to detect any reversal of *ciliary* currents by experiments with colouring-matter, and it is improbable that any such reversal occurs. I have, however, observed (*v. infrà*, p. 53) a violent reversion of currents, due, I believe, to suction, during the emission of *Glochidia*. This suction is probably effected by relaxation of the adductors and consequent partial opening of the shell while the right and left mantle-margins are kept in contact so as to block the aperature at all other parts except the two siphonal notches, of which the exhalant in particular remains open. The thickened margins of the mantle thus serve to temporarily close the aperture between the two valves, and, if my explanation be correct, the muscle-fibres of the mantle between the point of attachment of the mantle to the shell and its free border may tend to draw the right

and left thickened borders together in the middle ..., while ...
increasing their thickness and offering a more solid ....
water. Furthermore, when once the thickened borders of the ....
are in apposition and the shell commences to gape, the pressure on the
right and left free borders will tend to drive them even more closely
together; for the line of the mantle which is attached to the shell
must of necessity follow the outward movement of the valves when
.... commences, and the free borders unite to form a bluntly
.... longitudinal ridge with divergent sides; the pressure of
water falls on these divergent sides and drives them together—the
whole structure thus acting in the manner of the mitral valve of the
human heart. It is probable that the flexible margins of the valves
are also driven together by the pressure of water. The diagram
exhibited (Plate I. fig. 7) may make this clearer.

I am inclined to think, then, that a suction of this kind is used
to swiftly draw the ova forward into the external gill-plate. Direct
observation on this point is well nigh impossible owing to the necessity
of disturbing the animal or even partly opening the shell in order to
ascertain whether or no ova are in transit. The fact that violent
suction does take place in the case of the *Glochidia* is beyond doubt;
the exact mode of causing the suction is, for our present purpose, of
less importance.

The question naturally occurs, why do not the ova find their way
into the internal as well as the external gill? The reason is, I
think, twofold. In the first place, the space between the lamellæ
of the external gill is considerably greater than that between the
lamellæ of the internal gill. In the second place, as I have ascertained by careful dissection of many individuals, the inner lamell...
of each external gill-plate extends further towards the dorsal surface
than the outer lamel'a of each internal gill-plate, and stretches over
towards the middle line so as to greatly diminish or even totally
close the aperture leading into the space within the internal gill.
In some cases the inner lamellæ of the external gill-plates of the
right and left sides actually come in contact with one another ... the
median line posteriorly¹.

¹ This, of course, applies only to the posterior portion of the gill-plates.
In the region of the foot the .... close the space between
the lamellæ of the internal gill, as .... by von Baer.

The diagram (Plate I. fig. 8), which is a modification of Lankester's diagram (Encycl. Brit. 9th ed., Art. "Mollusca," fig. 135 D,
p. 690), will make these relations clearer.

II. *The Attachment of the* Glochidia *to the Parent Gill-plate.*

It is well known that the epithelium of the external gill-plate
secretes a nutritive mucus in which the young are imbedded and
thus retained within the gill.   This mode of attachment is, however,
not permanent; for if, as is often the case, the *Glochidia* are retained
for a long time after they have attained maturity, a large number
escape from their egg-capsules, and the so called "byssus," becoming
entangled in the gill-filaments and bars of concrescence, serves to
secure them until they are forcibly expelled from the parent.   I
have found that the number of *Glochidia* in any given parent which
have escaped from their egg-capsules varies with the period during
which they have been retained since the attainment of pre-parasitic
maturity.   It thus appears that as the nutritive mucus is used up,
and its power of retaining the *Glochidia* within the gill is therefore
diminished, a secondary mode of attachment becomes of all
importance and is furnished no longer by the parent but by the
adult members of the Glochidian family, in whose neighbourhood the
mucus has been chiefly absorbed and who alone are provided with
fully developed byssus-filaments.   This phenomenon is the more
interesting as furnishing yet another case of prolonged attachment to
the parent of the young of freshwater animals (*vide* Sollas, "On the
Origin of Freshwater Faunas," Scientific Transactions of the Royal
Dublin Society, vol. iv. ser. ii., 1886).

III. *Emission of* Glochidia.

The female *Anodon* is usually stated to retain the *Glochidia* within
the external gill-plates until fish are in the neighbourhood.   This is
certainly not always the case, for *Glochidia* were frequently emitted
in large masses and long cords after I had gently stirred the water in
which the Anodons were lying.   Schierholz ("Entwick. der
Unioniden," Denk. d. kais. Akad. d. Wiss. 1889, lv. pp. 183–214)
states that nodular ejection of *Glochidia* is abnormal, due to imperfect

aeration of the water and necessity of using the outer gill for
respiratory purposes, that normally ejection takes place singly with
the egg-capsules (cast off), which float off and leave the larvae in
_____ on the bottom. I fear I am unable to endorse this account
_____ ; nodular ejection undoubtedly is abnormal, but ejection in
_____ I have always found to occur in healthy individuals supplied
with well aerated water, and on one occasion have seen it occur in an
_____ *Anodon* it its native water. It would seem that any
_____ of _____ irrespective of fish, if not too violent, pro-
_____ of the *Glochidia* in a perfectly normal manner.

_____ important to notice that the parent is able *to draw back within
the shell the long slimy masses of* Glochidia even after they have been
_____ distance of 2 or 3 inches. The importance of this fact I
have already mentioned in dealing with the transit of ova. I
_____ the *Glochidia* on several occasions, in both *Anodon* and *Unio*,
_____ made "to enter a second time into their mother's womb."

IV. *Alleged Swimming of* Glochidia.

The belief that *Glochidia* can swim by clapping their valves together
"like *Pecten* or *Lima*" appears to be very general in this country, in
spite of frequent denials (*e.g.* Schierholz, *loc. cit.*). The extent of the
swimming-powers consists solely in "swimming to the bottom"; in
other words, *Glochidia cannot swim*. A *Glochidium* normally lies at
the bottom of the water on its dorsal surface, the ventral surface
being upwards and the "byssus" (so-called) streaming up into the
water above. In this position the *Glochidium* lies powerless to move
in any direction, and here, too, it dies unless a convenient "host" is
in some way brought in contact with its "byssus." If the water is
disturbed the *Glochidia* are carried about by currents, but soon fall to
the bottom again and are entirely unable to make headway in any
direction, even when they are thus temporarily suspended in mid-
water.

The *Glochidia* are evidently peculiarly sensitive to the odour (?) of
fish. The tail of a recently killed Stickleback thrust into a watch-
glass containing *Glochidia* throws them all into the wildest agitation
for a few seconds; the valves are violently closed and again opened
with astonishing rapidity for 15–25 seconds, and then the animals

appear exhausted and lie placid with widely gaping shells—unless
they chance to have closed upon any object in the water (*e.g.* another
Glochidium), in which case the valves remain firmly closed. I found
this excitement very useful in procuring *Glochidia* widely open.
Flooding with hot corrosive sublimate kills them instantly and the
shells remain apart.

V. *Relation of* Glochidium-*shell to Shell of Adult.*

So long ago as 1825 it was pointed out by Pfeiffer (Naturg.
deutscher Land- und Süsswasser-Mollusken, Weimar, 1825), and more
recently by Kobelt and Heynemann, that the shell of the *Glochidium*
sits like a saddle over the dorsal and lateral surfaces of the shell of
the young *Anodon* or may be seen in uninjured specimens close to the
hinge-line It has not, however, been noticed, so far as I can ascertain,
that this temporary situation of the *Glochidium*-shell has a permanent
effect upon the shape of the adult shell. This effect will be at once
apparent on referring to Plate I. figs. 2–5.

About 101 days after first attachment to the host and 25–30 days
after quitting the host, the shell-teeth of the *Glochidium*-shell project
ventrally towards the median line, and as a consequence impinge
upon the ventral border of the at present soft shell of the adult at
a point about halfway along its length, the result being that at this
point the permanent shell is prevented from growing so fast as elsewhere. The permanent shell at this stage, therefore, has its otherwise
symmetrical curve sharply interrupted by an irregular notch,
pointing towards the dorsal surface (*vide* figs. 2 & 3). *This notch,
in the vast majority of cases, persists through life* and causes a slight
dorsal turn of the curves marking the lines of growth at a point
roughly halfway along their length, but, as a rule, slightly nearer
the posterior border of the shell. In each successive line of the
growth the notch becomes of greater antero-posterior and less dorso-
ventral extent, thus tending to become less evident and to disappear.
The notch can therefore be seen most easily near the hinge-line
(*i.e.* on the first lines of growth) in those shells which have escaped
corrosion. In 15 species of *Unio* belonging to the collection of
Admiral Sir John Harvey in the University Museum, Oxford, this
notch is evident and undoubtedly caused in the way above described ;

it is perhaps present in 2 others (*V.* cylindricus and *U. triangularis*) and is quite clear also in 6 species of Anodon. The figures given by Chenu in his 'Manuel de Conchyliologie,' and by M. Henri Drouet, "Unionidæ du Bassin du Rhône," Mém. de l'Acad. des. Sci. Arts et Belles Lettres de Dijon, (4) i. 1888-89, pp. 27-113, pls. i.-iii., show the notch almost without exception. I do not rely strongly on these figures for this particular, as many irregularities of curvature occur, owing to individual injury at some period of life, and it is necessary to examine each specimen personally before deciding whether the notch figured can in every case be assigned to the Glochidian shell-teeth.

I may take this opportunity of corroborating Schierholz's statement (*loc. cit.*), concerning the absence of sexual distinction in the shape of the shell. It is commonly believed that the shell of the female is far more convex and of greater transverse diameter than that of the male. This is not the case : there is no point by which the shell of the female can be distinguished. On several occasions I have requested persons professing to be able to distinguish the sexes in this way to select a few males from my stock : out of 19 thus selected on various occasions only one proved on dissection to be of the male sex, whereas on one occasion a small *U. pictorum,* which was selected as "undoubtedly female" turned out to be a male ! I have invariably found males very rare and was long unable to procure one ; for instance, of 50 Anodons dredged from a small pond in Norfolk, and averaging between 3 and 4 inches in length, only two were males ; the same was true for Anodons and Unios collected out of the canal at Oxford, though here the proportion of males was slightly higher. So rare in fact were the males and so small were the majority of them, that I was tempted to believe that *Anodon* is hermaphrodite, functioning in early life as male and later as female ; I made several experiments to investigate this point, but obtained no evidence on either side. Stress of work has prevented me from making any further search in this direction.

VI. *The Cilia on the Foot of Young* Anodon.

While observing young Anodons of 3-6 weeks old (dating from the end of parasitic life), I was struck by the peculiar movements of

56 OSWALD H. LATTER, M.A.

the cilia covering the foot. While the animal is in motion the foot
is first protruded somewhat slowly until it stretches a considerable
distance in front of the anterior margin of the shell, the cilia all
the while moving with great rapidity and appearing to "feel the
way." The foot having been protruded to its utmost extent, the
shell is drawn forward by a rapid muscular contraction. As soon
as this contraction commences, the cilia suddenly cease moving and
stand out from the surface like the bristles of a brush absolutely
motionless and rigid. This condition is maintained until the foot
again commences to glide forward. I can offer no suggestion as
to the meaning or cause of this apparent rigidity other than that
the appearances are as though the pressure within the epithelial
cells becomes so great that the cilia cannot assume any other position
than one perpendicular to the surface, and forming a continuation or
the long axis of the cells on which they are severally carried.

VII. Glochidia *distasteful to Fish.*

All fish with which I have experimented, viz., Perch, Loach, Stickleback, Minnow, have a strong dislike for *Glochidia* as an article of
food. They frequently seize a mass of *Glochidia* floating in the
disturbed water, but the mass is no sooner within the mouth than it
is forcibly and emphatically rejected, being spit out to a considerable
distance and very rarely (only once) attempted again. I do not
think that it is the irritation caused by *Glochidia* attaching themselves to the inside of the mouth which makes the fish behave thus,
for I killed six fishes which had tasted *Glochidia* within ten minutes
of making the experiment, and in only one of them did I find a
Glochidium attached to the mouth. There must, I think, be some
unpleasant odour or taste about the *Glochidia*; or possibly the
"byssus," the shell-teeth, or both these latter combined, may
serve to make the *Glochidia* uninviting morsels.

VIII. *Powers of Resistance of Adult* Anodon *and* Glochidia.

An adult *Anodon* will live for at least a week, in cold weather,
after it has been removed from the shell. I consider the animal
alive so long as the cilia are beating and the heart is pulsating or
capable of responding to a moderate stimulus. The *Glochidia* will
live for a day or two within the gill of an apparently dead parent.

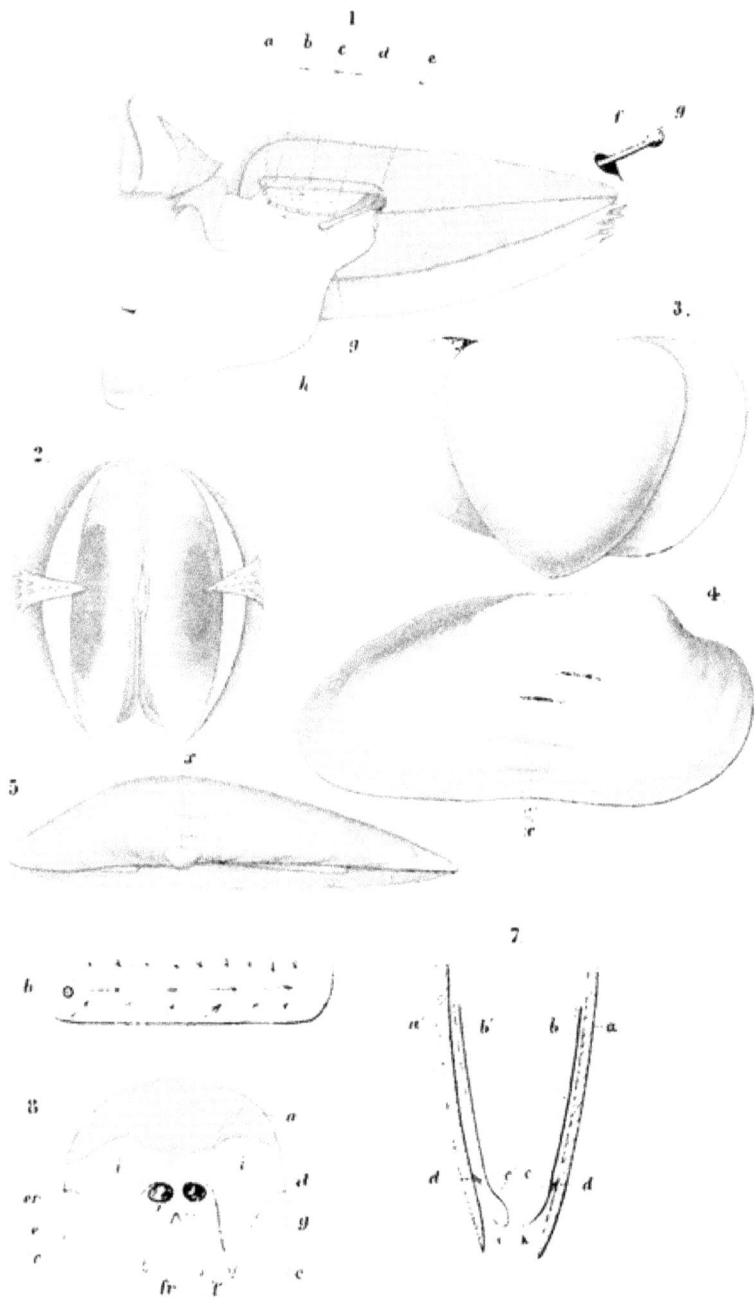

I was very much interested to notice one morning after a severe frost that the water in the dissecting-dish where an *Anodon* lay removed from its shell was completely frozen. I allowed the frozen mass to thaw gradually, and then examined the animal and its *Glochidia*; both were quite alive and none the worse for their severe exposure. I allowed the same animal and its young to be again frozen the following night, and obtained the same result. This power of being frozen and recovering must be of great importance in preserving the species in many of our shallower ponds and streams which are frequently frozen to the bottom in severe weather.

EXPLANATION OF PLATE I.

Fig. 1. Diagram of *Anodon* to show course of ova. The left mantle-flap has been reflected towards the dorsal surface and also the left gill-plates. The free dorsal margin of the inner lamina of the internal gill-plate has been turned up to show the surface of the nephridium (organ of Bojanus). *a*, external nephridial aperture; *b*, genital aperture; *c*, reflected free portion of dorsal margin of inner lamina; *d*, ciliated external surface of nephridium; *e*, retractor pedis muscle; *f*, exhalant siphonal notch; *gg*, probe passed through from lower to upper division of subpallial chamber, passing out at *f*; *h*, oviduct. The arrows indicate the direction in which the ova pass.

2. Ventral view of shell of young *Anodon*, 101 days after first attachment to host and about 25–30 days after the end of parasitic life. The *Glochidium* shell is shown outside the permanent shell, and the shell-teeth project inwards towards the middle line in such a way as to press upon and constrict the permanent shell at a point about half-way along its length.

3. Lateral view of somewhat old *Anodon*.

Figs. 4 & 5. Lateral and dorsal views respectively of left valve of small adult *Unio*, showing the notches *x*, produced on each line of growth by the previous constriction caused by the shell-teeth of the *Glochidium*-shell.

Fig. 6. Diagram to show the direction of ciliary currents on external surface of nephridium.

7. Diagram to show valvular action of ventral edge of mantle-flaps. *a*, *a'*, right and left valves of shell; *b*, *b'*, right and left mantle-folds; *c*, *c'*, thickened margins of *b*, *b'*; *d*, *d'*, lines of attachment of *b*, *b'* to *a*, *a'*. The arrows indicate the direction of water-pressure.

8. Diagram of relation of gill-lamellæ to show how the ova are prevented from falling into the internal gill. *a*, visceral mass; *c*, mantle-flap; *d*, axis of gill; *e*, inner, *er*, outer lamella of external gill-plate; *f*, outer, *fr*, inner lamella of internal gill-plate; *g*, line of concrescence; *i*, suprabranchial space of subpallial chamber.

Reprinted from the Journal of the Marine Biological Association, New Series, Vol. II., No. 1.

REPORT ON THE TUNICATA OF PLYMOUTH,

By Walter Garstang, *M.A., Fellow of Lincoln College, Oxford; formerly Berkeley Fellow of The Owens College, Manchester.*

With Plate II.

Part I.—CLAVELINIDÆ, PEROPHORIDÆ, DIAZONIDÆ.

The southern shores of the English Channel have long been famous for the wealth of their Tunicate fauna, having furnished material in abundance for the classical researches of Milne-Edwards, Giard, and Lacaze-Duthiers. The Channel Islands also have been repeatedly visited by English zoologists, and have amply supplied those among them who have been in search of Tunicate treasures. Probably the peculiar tidal conditions of this part of the Channel are especially favourable to a rich development of littoral forms; but, as the work of Montagu, Couch, Clark, Alder, Gosse, Cocks, Bate, and Norman sufficiently testifies, the Devon and Cornish coasts of England can lay claim to an almost equally luxuriant shore fauna, the rocky bays and long sheltered estuaries being especially wealthy in this respect. During my residence at Plymouth I found that the Tunicata were among the best represented groups of the fauna, and, as I devoted considerable attention to the search for rare or new, as well as for well-known forms, I trust that a classified report upon the local representatives of the group will not be without its usefulness to other investigators.

The absence of any work at all approaching the character of a monograph of the British Tunicata is a serious want which has long been felt by marine zoologists generally. Such a work has several times been commenced by some of our most eminent naturalists, by Forbes and Goodsir, by Alder and Hancock, and by Professor Huxley; but various causes have hitherto conspired to

delay its production.   It is now very satisfactory to be assured that
the preparation of the Monograph is in the hands of the experienced
author of the Reports on the " Challenger " Tunicata.   In the meantime a more or less detailed account of the forms with which I have
met at Plymouth may be of some service as a contribution towards an
improved knowledge of the British representatives of the group.

In the neighbourhood of Plymouth I found the rocks under the
Hoe, the north and east sides of Drake's Island, the wooden piles of
the docks and wharves in Millbay and the Cattewater, the rocks and
tidal pools of the Mewstone and Wembury Bay, to be all good
hunting-grounds for the littoral species of composite Tunicata ; the
best dredging-grounds for Ascidians generally were undoubtedly the
neighbourhood of the Duke Rock, the Queen's Grounds, and the
deeper waters off the Eddystone, the Mewstone, and Bigbury Bay,
while some forms were most common upon *Zostera* in Cawsand Bay ;
but it was almost impossible to use any of the ordinary methods of
collecting within Plymouth Sound without obtaining numbers of
Ascidians of various species.   Very few simple Ascidians were to be
found inhabiting the tidal zone ; they were most plentiful in the deep
water of the trawling grounds, on the rough ground off the Mewstone.

In reporting upon the Ascidians of Plymouth, I have taken
Clavelina and its allies as my starting point, since this genus includes
the forms which are in many respects probably the least modified
descendants of the earliest Ascidiacea.   But I am met at the outset
by the problem which is now engaging the attention of every
Ascidiologist : What taxonomical value must be attributed to the
possession of the power of budding and of the formation of colonies?
A full discussion of this question I cannot give here, but since the
matter bears directly upon the classification which I shall employ, I
am bound to admit that the division of the Ascidiacea into the sub-
orders *Ascidiæ simplices*, *Ascidiæ compositæ*, and *Ascidiæ salpiformes*
so completely disregards the admitted inter-relationship between
various sections of these groups, that its adoption seems to me to
involve the rejection of any morphological, and therefore genetic,
meaning in classification altogether.   The term " composite Ascidians " is in practice a very convenient one, but this is not a
sufficient reason for retaining it as the symbol of a natural group,

when the group in question is in reality no natural group at all, but an "artificial assemblage" composed of several quite unrelated phyla. The primary subdivision of the *Ascidiacea* into these three sub-orders will therefore not be adopted in my Report; the various genera will be grouped into families upon morphological grounds pure and simple, and will be taken as far as possible in the order of their affinity. From the nature of the case it is impossible to do this with perfect satisfaction, because the families of Tunicates, as of other order of animals, do not form a single series; but upon completing the description of the species I will present a scheme of classification in which the various families will be bound together according to their most probable phylogenetic relationships.

I desire here to express my warm thanks to Professor Herdman for the assistance which he has liberally given me from time to time ; as regards the present paper, I am particularly indebted to him for his kindness in rendering me various information concerning still unpublished work of his upon members of the *Clavelinidæ* and upon Tunicate classification* generally. I am equally indebted to Professor Milnes Marshall for the excellent facilities and help which he has afforded me in his Laboratories.

Order ASCIDIACEA.

Family 1.—CLAVELINIDÆ.

Body consisting of a thorax and abdomen connected by a slender, more or less elongate, œsophageal region. Stolonial tubes arising from the posterior end of the abdomen, rarely from its lateral walls.

Test gelatinous or cartilaginous; forming either distinct sheaths round the stolonial tubes or a common mass investing them ; common test never extending above the abdominal region ; apertures circular, not lobed, placed near together, terminal.

Musculature consisting almost exclusively of longitudinal bundles ; transverse muscles rare.

Branchial sac not folded ; horizontal membranes well developed,

* Professor Herdman's views upon the classification of the Tunicata will form the subject of a comprehensive memoir in the Transactions of the Linnæan Society, to which Society they were recently communicated.

without papillæ or internal longitudinal bars; dorsal lamina consisting of a series of languettes flattened antero-posteriorly and .... with the horizontal membranes; stigmata straight.

.... simple, filiform.

Genitalia in the loop of the intestine; oviduct and vas deferens present.

Reproduction by gemmation as well as from ova.

The family, as thus defined, includes the genera *Clavelina* (Savigny), *Podoclavella* (Herdman), *Stereoclavella* (Herdman), and a new genus, *Pycnoclavella*, described below.

I believe there is abundant reason for dividing the family Clavelinidæ, as regarded by Herdman, into several groups; *Perophora*, indeed, was excluded by Giard in 1872, and by von Drasche in 1883, while still more recently Lahille (1890) has emphasised the differences between *Clavelina* and *Perophora* by placing the former genus in the Distomidæ and the latter in the Ascidiidæ. Von Drasche's family Clavelinidæ includes *Diazona*, but although a near relationship between *Clavelina* and *Diazona* is generally admitted, it must be remembered that the new forms discovered in recent years have rather emphasised than reduced the gap between the two genera; I have therefore excluded *Diazona* from the Clavelinidæ altogether. As above defined, this family includes a number of forms about whose close mutual affinity there can be no doubt.

1. CLAVELINA, *Savigny.*

ASCIDIA, *O. F. Müller* Zoologia Danica, vol. ii, 1788. In part.

— *Bruguière.* Hist. Nat. des Vers, Encycl. Méthodique, Paris, 1792, p. 141. In part.

— *Turton.* Linné's General System of Nature, London, 1802, vol. iv, p. 92. In part.

CLAVELINA, *Savigny.* Mémoires sur les Animaux sans Vertèbres, Paris, 1816, IIe Partie, pp. 87 and 109. In part.

— *Savigny.* Tableau systématique des Ascidies, p. 171. In part.

— *Fleming.* Molluscous Animals, Edinburgh, 1837, p. 202. In part.

— *M. Edwards.* Observations sur les Ascidies des Côtes de la Manche Mém. de l'Acad. des Sci., Paris, xviii, 1842, p. 50. In part.

CLAVELLINA, *Alder.*   Cat. Moll. of Northumberland and Durham.
Trans. Tyneside Nat. Field Club, 1848, p. 108.

CLAVELINA, *Forbes and Hanley,*   Hist. Brit. Moll., i, 1853, p. 26.

CLAVELLINA, *Adams.*   Genera of Recent Mollusca, London, 1858,
vol. ii, p. 595. In part.

CLAVELINA, *Giard.*   Recherches sur les Synascidies, Arch. Zool. Exp.,
i, 1872, p. 613.

  —   *Herdman.*   Prelim. Report on Tunicata of "Challenger,"
Proc. Roy. Soc. Edin., x, 1880, p. 717. In part.

  —   *Herdman.*   Rep. on Tunicata, "Challenger" Reports.
vol. vi, p. xvii, p. 245. In part.

  —   *R. von Drasche.*   Die Synascidien der Bucht von Rovigno,
Wien, 1883, p. 8.

  —   *Carus.*   Prodromus Faun., Mediterr., Stuttgart, 1890,
vol. ii, pt. ii, p. 476.

  —   *Herdman.*   On the Genus Ecteinascidia and its Relations,
with Descriptions of two new Species, and a Classification of the Family Clavelinidæ, Trans. Biol. Soc.
Liverpool, vol. v, 1890, pp. 160, 161.

Body oblong, more or less clavate, not provided with a post-
abdominal peduncle.

Stolons distinct, delicate and branched.

Professor Herdman has quite recently* subdivided this genus as
it was defined in his "Challenger" Report. The difference recognised by Savigny between the types *borealis* and *lepadiformis*, in
regard to the presence or absence of a well-developed post-abdominal
peduncle of test-substance, is raised by him into a criterion of generic
value (*Podoclavella*), while the rudimentary "common test" enclosing
the stolons of the "Challenger" species constitutes the salient
character of his new genus *Stereoclavella.* These changes have been
rendered desirable by an increase in the number of species and the
apparent † distinctness of the three types. As the table in Professor Herdman's paper shows, the restricted genus now includes the
species *lepadiformis* (O. F. Müller), *rissoana* (M. Edwards; variety
only ?), *pumilio* (M. Edw.), *producta* (M. Edw.), *Savignyana* (M. Edw.),
and *nana* (Lahille).

   * *On the Genus Ecteinascidia,* &c., loc. cit.

   † One of the individuals in Milne-Edwards' figure of *Clavelina producta*
(l. c., pl. ii, fig. 3) exhibits a post-abdominal peduncle of some extent.

Is not Müller's *Ascidia gelatina* (Zool. Dan., iv, 26, plate 143) also probably a species of *Clavelina*?

1. CLAVELINA LEPADIFORMIS, *O. F. Müller.* (Pl. II., fig. 1).

ASCIDIA LEPADIFORMIS, *O. F. Müller.* L. c., p., 54, pl. lxxix, fig. 5.
— — *Bruguière.* L. c., pp. 142, 151, 152; pl. lxiii, fig. 10.
— — *Turton.* L. c., p. 95.
CLAVELINA LEPADIFORMIS, *Savigny.* Tableau systématique, p. 171.
— — *Fleming.* L. c., p. 202, pl. xvi, fig. 57.
— — *Milne-Edwards.* L. c., p. 267, pl. i, fig. i; ii. fig. 1.
— RISSOANA, *Milne-Edwards.* L. c., p. 267.
— LEPADIFORMIS, *Thompson.* Rep. on Fauna of Ireland, Rep. Brit. Ass., 1843, p. 264.
CLAVELLINA — *Alder.* L. c., p. 108.
CLAVELINA — *Forbes and Hanley.* L. c., p. 26, pl. E, fig. 1.
— — *Dickie.* Rep. on the Mar. Zoology of Strangford Lough, Brit. Ass. Rep., 1857, p. 105—111.
— — *Mennell.* Rep. of Dredging Expedition to Dogger Bank and Coasts of Northumberland, Trans. Tyneside Nat. Field Club, 1868, p. 12.
— — *A. M. Norman.* Last Rep. on Dredging among Shetland Isles, Rep. Brit. Assoc., 1868, p. 303.
— — *Giard.* Recherches, l. c., pp. 613—615, pl. xxi, figs. 2, 5; xxiii, fig. 2.
CLAVELINA LEPADIFORMIS, *McIntosh.* Marine Invertebrates and Fishes of St. Andrews, 1875, p. 54.
— — *Herdman.* Fauna of Liverpool Bay, vol. i, 1886, p. 296, and Proc. Biol. Soc. Liverpool, vol. iii, 1889, p. 245.
— — *Carus.* L. c., p. 476.

Colonies compact, zooids numerous.

Zooids more or less stout, moderately clavate, slightly compressed from side to side, without well-marked external differentiation into thoracic and post-thoracic regions; average height from a half to three-quarters of an inch.

Test gelatinous, perfectly hyaline and transparent.

Thorax one third of the total body-length; œsophageal region short; conspicuous opaque bands of a yellow, brown, or white colour mark the position of the dorsal, ventral and anterior peribranchial sinuses.

Branchial sac with about thirteen transverse rows of stigmata; horizontal membranes abroad

Habits.—Attached to rocks and stones (rarely to algæ and the backs of crabs, Müller) at the bottom of the tidal zone; seldom extending into 10 fathoms water.

At Plymouth fine colonies of this species have been found at extreme low water on the north side of Drake's Island, near the "Bridge" under Mount Edgcumbe Park, in tide-pools among the Renny Rocks, and in a few other localities. A few isolated zooids have also been dredged occasionally in 4 to 5 fathoms water near the Duke Rock, and, very rarely, in 10 to 15 fathoms off the Mewstone and Penlee.

On Pl. II., fig. 1, I have represented part of the series of languettes which extends along the dorsal median line of the branchial sac, in order to display their method of connection with the horizontal membranes. The languettes themselves are comparatively narrow and slender; they are compressed antero-posteriorly and are not connected with one another by any trace of a longitudinal lamina.

The horizontal (interserial or transverse) membranes are thin but well-developed, and may project sufficiently for each one to completely cover the row of stigmata immediately behind it. The free margins of these membranes are perfectly even; they are not in the least degree scalloped (festooned), and they show no trace of marginal papillæ.

2. PYCNOCLAVELLA, *gen. nov.*

Der.—πυκνος, closely united.

External appearance.—Zooids small and delicate, clavate, arising by slender stalks from a more or less thick, basilar mass of test-substance.

Body consisting of a small thorax, a slender, often greatly elongate œsophageal region, and a more dilated abdomen, the greater part of which is imbedded in the basilar mass of common test.

Test thin and delicate around the thorax, thicker and firm in the foot-stalks, dense and cartilaginous throughout the basilar mass; the latter is traversed in all directions by stolonial tubes, some of which even extend and branch in the œsophageal region of the zooids, where they remain sterile or, more rarely, give rise to new buds.

The partial imbedding of the posterior ends of the zooids in a basal mass of test is a character which is common to this genus and the genus *Stereoclavella*, as recently defined by Herdman (l. c., pp. 160, 161); but although this is the only character by which *Stereoclavella* has been as yet distinguished, a comparison of *Pycnoclavella aurilucens* with the described species of *Stereoclavella* shows that marked differences exist between the two genera. In *Pycnoclavella* the zooids arise by slender stalks from the common basal test, and there is a definite demarcation between the two regions; while in *Stereoclavella** it is almost impossible to speak of the common test as a distinct structure. The elegant and regularly clavate form of the free portions of the zooids, together with their delicacy and small size, are also points clearly separating the former genus from the species of *Stereoclavella*. It appears to me to be very probable that the chief character common to these two genera has been attained independently in each case, *Stereoclavella* having arisen from a species of *Clavelina* resembling *C. lepadiformis* in form and size, while *Pycnoclavella* is more akin to *C. producta*.

2. PYCNOCLAVELLA AURILUCENS, *sp. nov.*   (Pl. II, figs. 2 and 3.)

Colonies very variable in shape and size, as regards both the thickness and extent of the common test and the length of the free portions of the zooids.

Zooids with thorax slightly compressed from side to side, almost as broad as long, connected with the basal test by a slender cylindrical foot-stalk of varying length; thorax $\frac{1}{20}$ inch in length; foot-stalk from twice to ten times as long. Abdomen elongate, deeply embedded in the common basal test.

Colour.—A band of golden-yellow pigment extends along the ventral side of the thorax and is continued into the œsophageal region; it is absent from the dorsal side; this band gives a conspicuous colouration to the zooids, when seen alive with the naked eye.

* The preliminary description given by Professor Herdman of *S. australis* has no reference to the exact character of its common test.

E

Test of a pale green colour, semi-transparent; thin around the thorax, thicker and firm in the œsophageal region, cartilaginous in the basilar mass; traversed by stolonial tubes in the basal, abdominal, and even œsophageal regions; in the latter region (that of the foot-stalk) the tubes are generally sterile.

Apertures circular, proximate, in the median sagittal plane; branchial terminal, cloacal subterminal.

Branchial sac with seven to nine rows of stigmata; horizontal membranes well developed, broad; dorsal languettes borne on the horizontal membranes, long and stout; endostyle of great size; aperture of hypoganglionic gland simple, circular.

Cardiac structures (pericardium, epicardium) as in *Clavelina*; pericardium not recurved.

Habits.—Irregularly attached along with masses of Polyzoa (*Bugula, Scrupocellaria,* &c.), calcareous sponges (*Leucosolenia*), and compound Ascidians (*Botryllus, Didemnum*) to varied objects from rough ground in 10–20 fathoms water (*e.g.* cases of tubicolous Annelids, *Gorgonia* stems, shelly débris); rarely forming a thin carpet on the stems of red weeds, such as *Delesseria.*

I first noticed this beautiful little Tunicate in the winter and early spring of 1889, when it was dredged several times on rough ground off the Mewstone, on one occasion to the west of it, but generally from half to two miles south or south-west of the rock. This is certainly the best locality for the species at Plymouth, although curiously enough the first specimen was dredged on January 26th in shallower water inside the Breakwater, north-west of the chequered buoy. This first colony was attached to the stem of a *Delesseria,* and formed a thin crust over its surface, the zooids having very short stalks (see Pl. II, fig. 3); the colony was unusally free from adventitious foreign bodies, and the configuration of its parts, especially of the basal test, was much more obvious than in specimens dredged off the Mewstone. These latter colonies are almost inextricably bound up with Polyzoa, Botryllids, Sponges, and other organisms, forming tangled masses in which usually only the brightly gleaming heads of the zooids are visible, the basal test being hidden beneath numerous other organisms and foreign bodies. It is a very interesting fact that the stalks of the zooids are elongated in direct proportion to the abundance and height of the foreign organisms competing with them for space and oxygen, resembling in this respect numerous epiphytes and other vegetable growths in a thick Brazilian forest.

If the smaller zooids of a living colony be touched with a needle, the bright yellow thorax frequently withdraws itself completely from the greenish test of that region and disappears within the stalks or below the level of the corm. The larger zooids contract upon irritation, but do not completely withdraw in this way. On contraction they give, as it were, a stoop or bend towards the dorsal side—away from the side with the line of yellow pigment ; this is due to the fact that the longitudinal muscle-bundles are somewhat more numerous in the dorsal than in the ventral section of the body. Very rarely, if the irritation be continued, the larger zooids may also behave like the smaller ones.

Since I made these experiments with *Pycnoclavella* I have found that Forbes and Goodsir* noticed a precisely similar reaction in the case of their *Syntethys hebridicus*. Indeed, there are several interesting resemblances between the genera *Diazona* (*Syntethys*) and *Pycnoclavella*, the chief of which I may mention as being the greenish colour of the test and the embedding of the abdominal regions of the zooids in a thick basal mass of common test, the thoracic portions remaining free ; in *Diazona* this process is more complete than in *Pycnoclavella*.

With regard to the relations of this species to *Clavelina*, I have stated above that although there can be no doubt that both *Stereoclavella* and *Pycnoclavella* are closely allied to that genus, and, indeed, almost certainly derived from it, I believe others will agree with me that this species is more closely related to *Clavelina* itself than to the species of *Stereoclavella ;* its nearest ally seems to be Milne-Edwards' species *Clavelina producta.*† This species produces buds from the lateral walls of the abdomen as well as from the basal stolonial tubes, a fact hitherto without parallel in the Clavelinidæ. *Pycnoclavella aurilucens*, however, exhibits occasionally the same phenomenon (see Pl. II, fig. 2), and there can be little doubt that the stolonial tubes traversing the foot-stalks in this species, whether they remain sterile or produce buds, are the direct homologues of the fertile stolonial

* Trans. Roy. Soc. Edin., **xx**, 1853, p. 308.

† Milne-Edwards' Observations, l. c., p. 267, pl. ii, fig. 3.

tubes of the abdominal walls in *Clavelina producta*. The only other "social Ascidian" possessing morphologically similar structures is *Sluiteria rubricollis* (Van Beneden* and Sluiter), whose transparent test is traversed by several sterile stolonial tubes, branching dichotomously and terminating in a few delicate papillary prolongations on its surface.

These three species illustrate the probable manner in which the "vessels of the test" in Ascidiidæ arose phylogenetically ; at first few, short and completely fertile (*e.g. Clavelina producta*), they subsequently increased somewhat in number and extent, dividing dichotomously in the thickness of the test, and became less fertile (*e.g. Pycnoclavella aurilucens*) ; at a still later stage (represented by *Sluiteria rubricollis*) the tubes became completely sterile, and, though still not numerous, were essentially organs of the test. The loss of the power of blastogenesis altogether would now bring us to the stage occupied to-day by the species of *Ciona ;* while an increase in the number of the vessels would lead to the condition found in the greater number of simple Ascidians.

It is interesting to note also that these forms furnish confirmatory evidence of the view enunciated by Della Valle † that the sterile ectodermic tubes of the test have essentially a "palliogenic" function. In *Pycnoclavella aurilucens* the part of the test traversed by them is much thicker and firmer than the thoracic portion, and in *Sluiteria rubricollis* the test is, according to Van Beneden, thicker and more resistant than in *Ecteinascidia*. The test of "social Ascidians" generally is characteristically thin and soft, and this can be referred directly to the absence or very slight development of sterile "palliogenic" tubes. The softness and delicacy of the test of *Ciona* as compared with that of *Ascidia* is also a further confirmation of Della Valle's view.

A fully illustrated account of the anatomy of *Pycnoclavella* will appear in another journal later in the year, and with it will be published coloured sketches of the living colony.

* E. van Beneden, *Les genres Ecteinascidia, Rhopalæa et Sluiteria ; note pour servir à la classification des Tuniciers,* Bull. Acad. Roy. des Sci., &c., Bruxelles (iii), xiv, 1887, pp. 43, 44.

† See Arch. Zool. Exp., x. *Notes et Revue,* p. xli.

Family 2.—PEROPHORIDÆ.

Body undivided into thorax and abdomen ; viscera on the left side of the branchial sac.

Test transparent, for the most part thin and membranous, rarely traversed by a few sterile stolonial tubes ; never investing the stolons in a common basal sheath ; apertures generally well apart, the branchial terminal and the cloacal dorsal, lobed, or rarely proximate, terminal and only indistinctly lobed.

Musculature consisting almost exclusively of transverse fibres ; longitudinal fibres rarely present except around the apertures.

Branchial sac not folded, horizontal membranes absent or feebly developed, replaced or surmounted by interserial rows of papillæ ; papillæ simple and unbranched or supporting incomplete or complete internal longitudinal bars ; bars papillate or not papillate ; dorsal lamina a longitudinal membrane or represented by a series of slender languettes ; languettes rarely compressed from before backwards ; stigmata straight.

Tentacles simple, filiform.

Genitalia in the loop of the intestine ; oviduct and vas deferens present.

Reproduction by gemmation as well as from ova.

This family includes the genera *Perophora* (Wiegmann), *Perophoropsis* (Lahille), *Sluiteria* (E. van Beneden), and *Ecteinascidia* (Herdman, sens. strict.). In a complete system of classification it should be placed very near to Roule's group "Phallusidées," which embraces the genera *Rhodosoma*, *Ascidia*, and *Phallusia*.

A species of *Ecteinascidia* (*E. Moorei*), quite recently described by Herdman, appears from his figures to possess dorsal languettes flattened antero-posteriorly, and this is implied, though not directly stated, in the text of his paper. This condition of the languettes is unique within the family, and affords an approach towards the genera *Rhopalopsis*, *Rhopalæa*, &c.

3. PEROPHORA, *Wiegmann*.

> ASCIDIA, *Lister*. Some Observations on the Structure and Functions of Tubular and Cellular Polypi and of Ascidæ, Phil. Trans., pt. ii, 1834, pp. 378–382.

PEROPHORA, *Wiegmann*. Jahresbericht, Archives, 1835, p. 309.

ASCIDIA, *Fleming*. Molluscous Animals, Edin., 1837, p. 202.

PEROPHORA, *Forbes and Hanley*. Brit. Moll., 1853, p. 28.

— *Adams*. Genera of Mollusca, 1858, ii, p. 596.

— *Giard*. Recherches, l. c., p. 615.

— *R. von Drasche*. Die Synascidien, 1883, p. 8.

— *Herdman*. Tunicata, Encycl. Brit., 9th Edit.

— *Carus*. Prodr. Faun. Med., 1890, ii, pt. ii, p. 476.

— *Herdman*. On the Genus Ecteinascidia, l. c., p. 161.

Zooids quadrangular or oblong, rarely pyriform, never cylindrical, generally compressed from side to side.

Test thin, membranous, without sterile stolonial tubes ; apertures apart.

Branchial sac rarely provided with rudimentary horizontal membranes ; interserial papillæ triangular or tubular ; papillæ simple or each provided near its extremity with an anterior and posterior longitudinal process ; processes rarely fusing to form complete internal longitudinal bars ; dorsal lamina, a rudimentary or well-developed longitudinal membrane, supporting interserial languettes compressed from side to side.

Stigmata usually in four, rarely six, transverse rows.

Stolons delicate, distinct, creeping ; branches generally alternate in position.

The species included within this genus are at present four in number—*Listeri* (Weigmann), *Hutchinsoni* (Macdonald), *viridis* (Verrill), and *banyulensis* (Lahille). Of these, *P. banyulensis* may prove not to be distinct from *P. viridis*, as Herdman believes, while *P. Hutchinsoni*, despite Macdonald's careful description and figures, will probably be found on re-examination to present some structural characters not included in the above generic diagnosis.

In his recent paper on *Ecteinascidia* and its allies, Professor Herdman has anticipated me in a description of the interesting condition of the interserial papillæ in *P. viridis*. I can quite confirm his account by my observations on a number of specimens of a *Perophora* which Professor Weldon collected in the Bahamas and gave into my hands some time ago for description. Professor Herdman rightly interprets the bifid or trifid papillæ of *P. viridis* as "rudimentary or imperfect internal longitudinal bars," but so far, I believe, no perfect bars have been discerned in the branchial sac of *Perophora*. In some specimens, however, sent to me from the Zoological Station at Naples, and labelled "Perophora Listeri" I discovered some months ago that numerous perfect internal longitudinal bars actually existed, being supported upon the ends of flat triangular "connecting ducts" precisely as in *Rhopalopsis crassa* or *Ecteinascidia Moorei*, with this difference only, that small papillæ were frequently present at the points of junction. The existence of papillæ on the bars renders the affinity between *Perophora* and *Sluiteria* still closer than has been already believed. It is very probable that a new species must be created for the Naples type, but that is a matter to which I hope to refer in a subsequent paper on the anatomy and variation of the genus. (See *Postcript*, p. 81).

3. PEROPHORA LISTERI, *Wiegmann*. (Pl. II, figs. 4, 5, 6).

    ASCIDIA, sp., *Lister*. Phil. Trans., 1834, pp. 378—382, pl. xi.

    PEROPHORA LISTERI, *Wiegmann*. Archives, 1835, p. 309.

    ASCIDIA, n. sp., *Fleming*. Moll. Anim., 1837, pp. 202—209, pl. xvii, fig. 59 (2).

    PEROPHORA LISTERI, *Forbes and Hanley*. L. c., p. 28, pl. E, fig. 2.

    —   —   *Giard*. Recherches, l. c., pp. 615, 616, pl. xxi, figs. 3, 6 to 11, 13 to 15, pl. xxiv.

    —   —   *Herdman*. Second Report, Proc. Biol. Soc. Liverpool, iii, 1889, p. 246.

    —   —   *Herdman*. On the Genus Ecteinascidia, l. c., pp. 158—161.

Zooids quadrangular, compressed from side to side, colourless, transparent.

Apertures widely separated, branchial with six lobes, cloacal with five.

Tentacles forty in number, of three sizes.

Branchial sac always provided with unbranched digitiform or slightly triangular interserial papillæ; no rudiments of internal longitudinal bars; rudimentary horizontal membranes; stigmata in four transverse rows, two between each pair of interserial papillæ.

Musculature feebly developed; transverse fibres few, widely separate from one another, extending from the dorsal region to the middle of each side ; also forming a weak sphincter round each aperture ; longitudinal fibres almost as well developed as the transverse, extending from the oral sphincter as far as the level of the first interserial bar of the branchial sac ; several longitudinal fibres arising anteriorly between the oral aperture and the anterior end of the endostyle, extending with the longitudinal fibres of the oral sphincter to the same distance ; longitudinal fibres of the cloacal sphincter short.

Habits.—Attached to stones or algæ in shallow water.

At Plymouth *Perophora Listeri* has been dredged in the estuary of the Yealm, and in 4 to 5 fathoms water off the Duke Rock. Mr. Heape recorded it as abundant on the rocks below the Hoe.

There can be very little doubt that the name given by Wiegmann to Lister's *Perophora* has been also applied to forms specifically distinct from it. Lister, in his admirable paper, remarks upon the existence of "finger-like processes, about eight in a row, that project nearly at right angles into the central cavity" [of the branchial sac], and these are shown in some of his figures.

Giard also mentions these papillæ and compares them with the papillæ which were figured by Savigny in his account of *Diazona violacea*. These papillæ are simple and digitiform; so that Giard's species probably did not differ from Lister's with respect to these structures.

On the other hand the species found at Naples and, as I gather from Professor Herdman's paper, at Banyuls also (by Lahille) present considerable differences from this simple arrangement. It is probable, therefore, that *Perophora Listeri* does not occur in the Mediterranean but is confined to the Atlantic shores of Northern Europe.

The condition of the papillæ in Plymouth specimens is shown on Plate 11, fig. 6, in a drawing taken from preserved material. These structures are seen to have a flattened triangular shape and are connected at their bases by very low and rudimentary horizontal membranes (cf. fig. 7). In life, these papillæ assume a more extended digitiform shape, as Lister long ago stated. If these papillæ were to be connected by internal longitudinal bars (as frequently occurs in the Naples species), meshes would be formed, each containing two stigmata.

The opening of the duct of the hypoganglionic gland (fig. 7, *c. v.*) is simply circular. It is situated in front of a raised triangular area, whose apex is posterior; this constitutes what is undoubtedly the homologue of the epipharyngeal groove. A precisely similar structure has been figured by Roule for *Rhopalaea nepolitana*,[*] and observed in *Sluiteria rubricollis* by E. van Beneden.[†] From the posterior apex of this area arises the dorsal lamina (fig. 7, *d. l.*) as a low membrane which increases slightly in height as it extends posteriorly At the level of each horizontal membrane it rises up into a curved triangular languette (*l*), and occasionally there is a small projection from its edge between each pair of interserial languettes (fig. 8, *i. p.*). An examination of fig. 7 also shows that the horizontal membranes really are continued upon the lateral faces of the dorsal lamina, although they do not extend along the languettes.

The structure of the dorsal lamina in this species approaches closely in essential features that described by van Beneden in *Sluiteria rubricollis*, in which form there is a continuous longitudinal membrane whose border is cut into festoons in correspondence with the number of transverse (interserial) bars. The lamina is provided with fourteen oblique ridges which also correspond in number with the horizontal bars. Although in his diagnosis of the genus *Sluiteria*, Professor van Beneden denies the presence of horizontal membranes (*l. c.*, p. 43), he admits in his description of *S. rubricollis* that the connecting ducts of the internal longitudinal bars "spring by an enlarged base from little interserial folds traversing the length of the transverse bars" (p. 34). This is precisely the condition I have found in the Naples *Perophora*, and it is essentially similar to what is here described for *P. Listeri*; interserial membranes are in each case present, but rudimentary. The ridges on the lamina of *S. rubricollis* are therefore undoubtedly of the same nature as the less conspicuous elevations formed in *P. Listeri* by the continuation of the horizontal membranes upon the sides of the lamina (see fig. 7). Further the lamina of *Slueteria rubricollis* is described as being enrolled, the concavity being to the right; my figure also shows that the marginal languettes of *P. Listeri* are bent over in a precisely similar way.

• Roule, *Rev. des Esp. de Phallusiadées de Provence*, Rec. Zool. Suisse, iii, pl. xiv, fig. 14.

† Ed. van Beneden, Sur les genres Ecteinascidia, &c., l. c., p. 35.

In *Ecteinascidia turbinata* (Herdman) and *diaphanis* (Sluiter) the dorsal lamina is represented by tentacular languettes unconnected by a longitudinal membrane. This membrane is present in *E. Thurstoni* (Herdman), while horizontal membranes are quite absent, and with them also every trace of interserial ridges on the sides of the dorsal lamina. In *E. Moorei* (Herdman) all the horizontal structures are well developed, but the longitudinal lamina is absent.

Family 3.—DIAZONIDÆ.

Body large, consisting of a thorax and abdomen connected by a slender, more or less elongate œsophageal region; stolonial tubes arising from the posterior end of the abdomen.

Test gelatinous or semi-cartilaginous, greatly developed around the basal stolonial tubes, with formation of a thick common test, in which the abdominal portions or the entire bodies of the zooids are imbedded; apertures terminal, each divided into six lobes, rarely smooth.

Musculature consisting of both longitudinal and transverse fibres, which for the most part anastomose freely; longitudinal fibres especially well developed.

Branchial sac large; with or without festooned horizontal membranes; interserial papillæ always present, supporting complete or rudimentary internal longitudinal bars; longitudinal bars not papillate; dorsal lamina represented by a series of languettes with long tapering ends; dorsal tubercle a large, longitudinally ovate slit surrounded by broad raised margins; branchial sac not folded.

Heart recurved upon itself.*

Genitalia in the loop of the intestine, or extending considerably behind it; oviduct present or absent.

Reproduction by gemmation as well as from ova, with formation of colonies of great size; colonies without systems.

This family, including the genera *Diazona* (Savigny) and *Tylobranchion* (Herdman), has relations both with the Cionidæ, Distomidæ, and Polyclinidæ. To Lahille belongs, I believe, the credit of first emphasizing

* This has not yet been established for *Tylobranchion*, but is probably the case.

the resemblances between *Diazona* and *Tylobranchion*, the latter being
one of the most interesting of the "Challenger" forms make known to
us by Professor Herdman's researches. As I gather from Prof. Herdman's remarks upon Lahille's system of classification, this zoologist
groups together, along with *Diazona* and *Tylobranchion*, the genera
Ecteinascidia, *Rhopalaea*, and *Ciona*. I believe, however, that *Ecteinascidia* is much more closely related to *Sluiteria* than to any of these
genera, and while Roule has established the relationship of *Rhopalaea*
and *Ciona* beyond doubt, the equally close affinity of *Diazona* to these
forms is still a matter of some uncertainty. That the mere formation
of a huge common mass of test enclosing the abdominal regions of the
zooids is not of itself a point of great systematic importance is
demonstrated by *Pycnoclavella*, which is in every other structural
respect a true *Clavelina*. Therefore on this head I am quite in accord
with Lahille in his efforts to break up the group "Ascidiæ compositæ,"
and to classify the Ascidians upon morphological grounds only or in the
main. Yet I cannot but regard the definite position of *Diazona* and
Tylobranchion among the Cionidæ as too forcible a disregard of the ties
which also bind them to the Polyclinidæ and Distomidæ.

4. DIAZONA,* *Savigny.*

 DIAZONA, *Savigny.* Tableau systématique, l. c., pp. 174, 175.
  — *Dujardin*, in *Lamarck.* Hist. Nat. des Anim. sans Vertebres,
   2nd ed. (par Deshayes and M. Edwards), t. iii, 1840,
   pp. 498, 499.
 SYNTETHYS, *Forbes and Goodsir.* On some Remarkable Marine
   Invertebrata new to British Seas, Trans. Roy. Soc.
   Edin., xx, 1853, p. 307.
  — *Forbes and Hanley.* Brit. Moll, iv, p. 214.
 DIAZONA, *Alder.* Observations on British Tunicata, Ann. and Mag.
   of Nat. Hist. (iii), xi, 1863, p. 169.
  — *Lahille.* Comptes Rendus, cii, 1886, p. 1573, and civ, 1887,
   p 210.
  — *Giard.* Comptes Rendus, ciii, 1886, p. 755, 756.
  — *R. von Drasche.* Die Synascidien, p. 8.
  — *Carus.* Prodr. Faun. Med., 2, ii. p. 180.

* I greatly regret that my efforts to obtain a copy of Della Valle's Con-
tribuzioni have been unsuccessful up to the time of going to press, and I
must express the same regret with regard to Lahille's Recherches sur les
Tuniciers.

Colony gelatinous, sessile ; the zooids superior, their thoracic portions freely projecting.

Musculature with longitudinal fibres united into well marked bundles.

Branchial sac with festooned horizontal membranes, supporting complete, rarely incomplete, internal longitudinal bars, not papillate at the point of junction ; dorsal languettes triangular, compressed from before backwards.

Genitalia in the loop of the intestine ; oviduct (always ?) and vas deferens present.

In 1853 Professors Forbes and Goodsir announced the discovery of a composite Tunicate allied to Savigny's *Diazona violacea*, but differing from it in the possession of the following characters : Plain undivided orifices, non-pedunculated abdomen, meshes with "one ciliated opening" only, and apple-green colour. Their genus *Syntethys* was established upon these grounds, but Alder subsequently wrote to show the generic identity of the two forms, basing his criticisms upon an examination of specimens dredged near Guernsey and possibly upon a re-examination of a portion of one of the original specimens of Forbes and Goodsir. Alder satisfactorily showed that the difference of colour was one due entirely to the action of the spirit used in preservation, and also that the pedunculation of the abdomen is very variable in its extent. He also noted that, after preservation, the division of the apertures into lobes was generally difficult to make out. His conclusion in regard to the generic identity of *Diazona violacea* and the so-called *Syntethys* is probably correct, although, as I shall endeavour to show below, his identification of the Guernsey species with that of Forbes and Goodsir from the Hebrides is extremely doubtful.

4. DIAZONA VIOLACEA, *Savigny.* (Pl. II., figs. 7, 8.)

DIAZONA VIOLACEA, *Savigny.* Mémoires, pp. 35—38, 175, 176, pl. xii.

    —   —   *Fleming.* Moll. Animals, 1837, p. 211.

    —   MEDITERRANEA, *Dujardin.* L. c., pp. 499, 500.

    —   HEBRIDICA, *Ald. c.* L. c., p. 169.

    —   VIOLACEA, *Carus.* L. c., pp. 480, 481.

Colony massive, irregularly rounded, attached by a short, thick pedicle or base ; total diameter about 7 inches, total height 5 or 6 inches : of apple-green colour when alive, semi-transparent.

Zooids often 2 inches long, with oral and cloacal orifices each six-rayed.

Branchial sac with sixty to eighty transverse rows of stigmata ; meshes each containing three, rarely four stigmata ; internal longitudinal bars for the most part completely formed ; but here and there represented by T-shaped interserial papillæ, as in *Tylobranchion ;* dorsal tubercle a large deep groove, elongate antero-posteriorly, with thickened walls.

Habits.—Attached to rocks and stones in deep water.

Dredged at Plymouth on rough ground off Stoke Point, and off the Eddystone in 20–40 fathoms of water.

There are two remarkable statements in the original description of the structure of *Syntethys Hebridicus* by Forbes and Goodsir which have not, to my knowledge, received the attention which they deserve. They are involved in the following account given by these naturalists of the branchial sac in their specimens :

" Branchial chamber with thirteen transverse rows of oblong openings, fringed with ciliated epithelium ; hooked fleshy tubercles at the intersections of the branchial meshes, each mesh presenting one of the ciliated openings ; the tubercles give the internal surface of the chamber a dotted appearance." (Trans. Roy. Soc. Edin , 1853, p. 307, cf. also for Forbes and Hanley, l. c., p. 244.)

Now, in the specimens of *Diazona violacea* dredged at Plymouth, the number of transverse rows of stigmata greatly exceeds that given by the eminent naturalists who described *Syntethys Hebridicus ;* the number is usually about sixty, seventy, or even more ! Further, the stigma in each mesh are invariably three or four, the latter number agreeing with the description and figure given by Savigny.

Were Professors Forbes and Goodsir mistaken ? Such a theory is unlikely, for one of their figures (l. c., pl. ix, fig. 4 *d*) shows in outline some of the appearances which they recorded in the words quoted above. Indeed, this figure is too precise to admit of any doubt as regards the approximate number of *transverse* bars (and, therefore, rows of stigmata) in their specimens, and a difference in this respect

between *Diazona violacea* and *Syntethys Hebridicus* must, I think, be
admitted.

But the more remarkable statement is that "each mesh presents
one of the ciliated openings." That Forbes and Goodsir should have
made a mistake in the observations which gave rise to this statement
seems inconceivable, but it is surprising that they pass no reflection
upon so unusual a condition of the brancial sac. There was plainly
no error in the identification of the "meshes," for "hooked fleshy
tubercles" are stated to be present at the "intersections of the
branchial meshes" (a somewhat confused but quite intelligible statement). Still, the fact of one stigma alone being included in each mesh
has either to be accepted or explained away.[*]

It is conceivable that the appearance of one "ciliated opening"
corresponding to each mesh was due to a great transparency of the
"trame fondamentale" of the branchial sac, and that while the meshes
were observed, the true stigmata were not noticed; but I cannot
reconcile this hypothesis with the assertion, so definitely made, that
the "oblong openings" were "fringed with ciliated epithelium."
It is almost impossible, and for the same reason, to imagine that the
meshes were totally devoid of stigmata, as in *Pharyngodictyon
mirabile* of the "Challenger" collection, described by Herdman.[†]

I am obliged, therefore, to conclude that *Syntethys Hebridicus*
actually possessed, as Forbes and Goodsir stated it to possess, a
branchial sac containing about thirteen transverse rows of oblong
stigmata, and presenting a "hooked fleshy tubercle" at the junction
of every longitudinal and horizontal bar.

It should be noticed that in the original description there is
nothing irreconcilable with the view that the branchial sac of
Syntethys Hebridicus may in reality have been quite destitute of true
internal longitudinal bars, and possibly of horizontal membranes;
the "hooked fleshy tubercles" may have been such rudimentary connecting ducts and bars as Herdman has described and figured for
Tylobranchion speciosum (l. c., p. 161). In this connection I may
state that I find the internal longitudinal bars of *Diazona violacea* to

---

[*] This condition exists in *Polyclinum sabulosum* (Lahille, Comptes Rendus,
cii, p. 1574), and is approached in *Tylobranchion speciosum*.

[†] Herdman, "Challenger" Report, vol. xiv, pt. xxxviii, p. 155.

be by no means rarely incomplete in portions of the branchial sac ;
they are then represented by structures which could well be described
as " hooked fleshy tubercles."

I will not maintain that this new view of Forbes and Goodsir's
very "remarkable invertebrate" is probable, but it is at least
possible. If it should prove eventually to be correct, a very interesting connection between *Diazona violacea* and *Tylobranchion
speciosum* will have been established.

By admitting the above-named differences between the branchial
sacs of *Diazona violacea* and *Syntethys Hebridicus*, it will be noticed
that I do not accept Alder's identification of his Guernsey specimens
of *Diazona* with Forbes and Goodsir's species. From Alder's account
I have been led to believe that he assumed this identity too hastily.
He states that his specimens were "at once recognised as the *Syntethys
Hebridicus* of Forbes and Goodsir," and upon this assumption he
endeavoured to find out what structural differences there might be
between this form and the *Diazona violacea* so admirably described by
the great French anatomist. His researches were not very fruitful of
result : "The only difference I can find is that the papillæ of the
branchial sac in the latter (*Syntethys Hebridicus*) are stout and obtuse,
very different from the slender pointed form represented by Savigny ;
I have therefore determined to consider them distinct until further
observations decide the point."

Now Alder's Guernsey specimens are certainly identical (specifically)
with the forms investigated by myself, and they are both from practically the same region of the English Channel ; there is further no
appreciable difference between the Plymouth forms and Savigny's
species. Therefore Alder's examples must also be referred to the
species *Diazona violacea*.

The absence of any indication in Alder's paper that he re-examined
the "portion of a specimen (of *Syntethys Hebridicus*) from the
original habitat" which Professor Goodsir sent to him, renders intelligible what would otherwise have been a very strange omission on
his part. I refer to his failure to notice the remarkable discrepancy
between the structure of the branchial sac in the Channel specimens
and that described for *Syntethys Hebridicus*.

Thus, although I quite agree with Alder that there is as yet no

sufficient ground for generically separating these two forms, I believe
Forbes and Goodsir's species to be perfectly distinct, and to possess
the following distinguishing characters :

Orifices not divided into lobes, evenly rounded.

Zooids projecting freely by their thoracic portions, which are
united to the common mass by a slightly contracted but not pedun-
culated œsophageal region.

Branchial sac with thirteen transverse rows of oblong stigmata ;
a hooked interserial papilla (connecting duct) at the intersection of
every longitudinal (interstigmatic) and transverse bar.

Oviduct absent (?)

But the whole matter is so beset with doubts that it is greatly
to be desired that specimens should be obtained again from the
Hebrides, and their anatomy re-described. Unfortunately Giard gives
no anatomical details of the *Diazona* dredged by him and described
in Comptes Rendus, ciii, p. 755.

DESCRIPTION OF PLATE II.

Illustrating Mr. W. Garstang's "Report on the Plymouth Tunicata,"
pt. 1.

N.B.—Where not otherwise stated all the figures were drawn from preserved
material.

FIG. 1.—*Clavelina lepadiformis*, O. F. Müller. Three of the dorsal languettes
of the branchial sac. Zeiss, A, Oc. 2, Cam.

    d.s. = Dorsal sinue.

    h.m = Horizontal membrane.

FIG. 2.—*Pycnoclavella aurilucens*, gen. et sp. nov. Portion of a colony. A
section has been taken through the colony in order to show the common test
and the imbedded portion of the zooids. Nat. size ; slightly diagrammatic.

    c.t. = Common test.

    t.p. = The free portions (thoracic and œsophageal) of the zooids pro-
jecting from the common test.

    v.p. = The visceral portions of the zooids imbedded in the common test.

FIG. 3.—*Pycnoclavella aurilucens*, gen. et sp. nov. The free portion of a
zooid from a colony growing on the stem of a *Delesseria*. Drawn from life,
enlarged.

FIG. 4.—*Perophora Listeri*, Wiegmann, Portion of the branchial sac. Zeiss,
A, Oc. 2, Cam. luc.

    h.m. = Rudimentary horizontal membranes.

    p. = Interserial papillæ (= rudimentary connecting ducts).

Fig. 2.

Fig. 1.

n s

v p

p

Fig. 3.

p p

J

h m

h m

Fig. 5.

v

Fig. 4.

p

p

Fig. 3

h m

Fig. 6.

l b

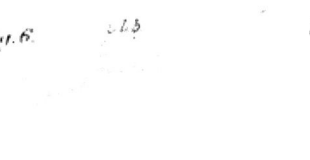

Fig. 7.

l b

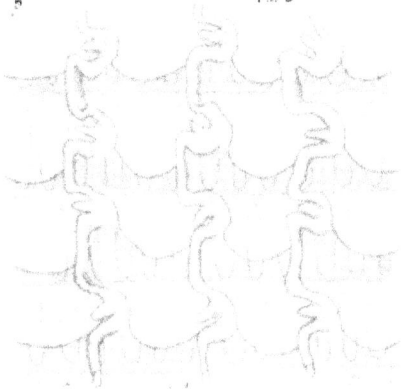

p

Fig. 8.

m

h m

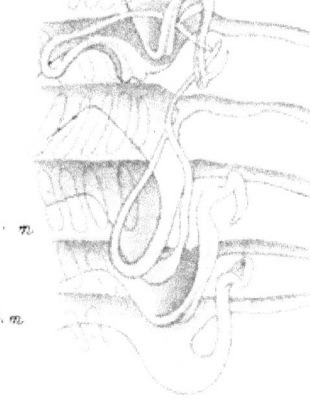

Fɪɢ. 5.—*Perophora Listeri*, Wiegmann. Dorsal wall of pharynx, showing dorsal lamina and aperture of hypoganglionic gland, seen from inside. Zeiss, A, Oc. 2, Cam. luc.

   p.g. = Pericoronal groove.

   c.v. = Ciliated vesicle, opening on the surface of a shield-shaped pad.

   g. = Ganglionic mass.

   d.l. = Longitudinal membrane of dorsa lamina.

   l. = Marginal languettes.

   h.m. = Rudimentary horizontal membranes.

Fɪɢ. 6.—*Perophora Listeri*, Wiegmann. Dorsal lamina of another individual, seen from the right side. Zeiss, A, Oc. 2, Cam. luc.

   l. = Marginal languettes, interserial in position.

   i.p. = Small marginal projections intermediate between the languettes.

Fɪɢ. 7.—*Diazona violacea*, Savigny. Portion of branchial sac, seen from inside. Magnified, slightly diagrammatic.

   h.m. = Horizontal membranes.

   i.l.b. = Internal longitudinal bars.

   p. = Papillæ of the connecting ducts. (See *Postscript*)

Fɪɢ. 8.—*Diazona violacea*, Savigny. Six dorsal languettes. Zeiss, A, Oc. 2, Cam. luc

   h.m. = Horizontal membranes.

POSTSCRIPT.

By Professor Herdman's kindness I have recently been enabled to consult Lahille's important *Recherches sur les Tuniciers*. Lahille points out that the appearance of papillæ on the internal longitudinal bars of the branchial sac of *Diazona violacea*, as previously described by Savigny and Della Valle, is a false one, produced by the thickened remains of the "primitive branchial languettes." I had myself, like Alder, failed to find any such vertical papillæ as were represented by Savigny for this species, and was struck by their apparently recumbent position in mounted preparations (see fig. 7); but a re-examination by means of dissecting needles has convinced me that Lahille is quite correct in denying their existence altogether. The necessary correction has been made in the text of my paper, but the diagram given on fig. 7 is in this respect misleading. Lahille also states that the horizontal membranes are very little developed, but this is by no means the case in the Plymouth specimens.

I have stated above (p. 71) that my discovery of internal longitudinal bars in the Naples *Perophora* will probably necessitate the creation of a new species; but from Lahille's description this species would appear to be identical with his *P. banyulensis*. The Naples species differs widely from *P. viridis* as regards its musculature, a fact which thus militates against Professor Herdman's suggestion that *P. banyulensis* is a synonym of *P. viridis*.

W. G.

Reprinted from the Quarterly Journal of 'Microscopical Science' for October,
1891.

THE FORMATION AND FATE OF THE PRIMITIVE STREAK, WITH OBSERVATIONS ON THE ARCHENTERON AND GERMINAL LAYERS OF RANA TEMPORARIA.

By Arthur Robinson, M.D., *Senior Demonstrator of Anatomy in the Owens College; and* Richard Assheton, M.A., *late Demonstrator of Zoology in the Owens College.*

With Plates III. & IV.

Even a superficial perusal of the literature of the development of the Amphibian ovum is sufficient to acquaint the inquirer with numerous contradictory statements concerning points of weighty morphological importance. Not only are there differences of opinion as to the interpretation of some of the developmental features which which are generally allowed to be readily recognisable, but there are also assertions, made by different observers, concerning merely the structural arrangement and the fate of various portions of the Amphibian ovum which are absolutely irreconcilable with each other. The confusion is increased by a loose application of terms describing changes and areas, so that it becomes difficult to decide the exact relation that the Amphibian ovum bears to the ovum of other Vertebrates.

In view of the chaos which already exists we would not readily enter upon any further discussion of Amphibian development, were it not that we think the results of our recent observations on the development of the Anura tend to simplify the problem we have been studying by throwing light upon several interesting developmental phenomena. For these reasons alone are we induced to publish the results of a research which tend to conclusions different from those of our predecessors, whose experience is, in most cases, much more extensive than our own.

At the outset it may, with advantage, be stated that our observations were directed principally to the mode of formation of the archenteron and blastopore, the fate of the blastopore, and the formation and fate of the primitive streak; but incidentally we have dealt with the separation of the germinal layers and the relation of the mesoblast to the chorda and hypoblast.

We worked independently of each other. One of us was led on to the investigation by the contradictions between the current statements regarding and the obvious facts disclosed by the developing anural ovum; whilst the other approached the subject with the conviction, based upon theoretical grounds, that either the descriptions of certain phases of Amphibian development were incorrect, or that there were very peculiar and significant differences between the Amphibian and other Vertebrate ova. Until our conclusions had been arrived at neither of us was aware that the other was engaged upon the subject. Consequently our observations were made and our conclusions formed independently of each other.

Methods.—Living and hardened ova were examined, but we relied chiefly upon sections cut in the three usual planes, horizontally, sagitally, and transversely; and of these the horizontal have proved in some respects the most useful and instructive, more especially in the observations upon the fate of the primitive streak.

The ova were hardened either in Perenyi's fluid or in Kleinenberg's picro-sulphuric acid solution. The former fluid was most useful in the younger stages. After the hardening was completed the ova intended for sections were embedded in paraffin in the usual manner.

None of the ova were stained, for our experience has been that the solutions through which the ova are passed, before the staining is complete, alters the relations of the cells, and that the staining hinders rather than facilitates the microscopical examination of sections of the younger stages.

We obtained several surface views which were useful for comparison with the results of previous observers; of these the one represented in fig. 11 was drawn from a living embryo.

Fig. 23 was drawn partly from a surface view of an ovum hardened in Kleinenberg's fluid, and partly from a model constructed by putting together, in order of sequence, pieces of cardboard cut to correspond with camera drawings of a complete series of transverse sections. Our work was greatly facilitated by Professor Marshall, who very kindly placed at our disposal a large series of sections, and to him also our thanks are due for much kind advice.

Before proceeding to the description of our own observations, we must refer shortly to previous records, for by this course alone shall we be able to indicate clearly the differences between our results and those of the observers who have preceded us, and at the same time we shall obtain an opportunity of defining some of the terms that we shall be obliged to use in our description.

Turning, therefore, in the first place to the consideration of the formation of the archenteron and the blastopore (but leaving aside for the present the concrescence theory of His [22 and 23], so far as it concerns the formation of this cavity), we find that upon this, as upon most other important points, there is a distinct difference of opinion between the previous observers. It is stated by some that after the formation of the segmentation cavity, and when the ovum is only partially covered by the pigmented epiblast-cells, the archenteron is formed by an invagination, which "first commences by an inflection of the epiblast-cells for a small arc on the equatorial line which marks the junction between the epiblastic cells and the yolk-cells" (1, p. 102).

This preliminary invagination is also described by Perenyi in Bombinator (40), O. Schultze in *Rana fusca* (45), and by Scott and Osborn in the Newt (48).

Whilst it is proceeding the enclosure of the yolk has been rapidly taking place. "It is effected by the epiblast growing over the yolk at all points of its circumference" (1, p. 102). We have in this latter statement of Balfour's a very fair summary of the general opinion that the epiblast and yolk are distinct parts of the ovum; indeed, Balfour speaks of the yolk-cells as "those large cells which are part of the primitive hypoblast" (1, p. 101), with which, in our opinion, they cannot fairly be compared; and we shall speak of them in the sequel merely as yolk-cells, though probably the more correct term would be germ segments. It is not necessary, however, to enter into a discussion of the mode of extension of the epiblast at the present moment, and we can proceed, therefore, to a point which is allowed by all, namely, that the superficial extension of the epiblast terminates at the margin of a large circular opening, called the blastopore or the anus of Rusconi, at the margin of which the surface epiblast becomes continuous with the cells lining the archenteric cavity, which, on account of their position, are spoken of as hypoblast, although those situated in the dorsal wall of the cavity are said to be invaginated epiblast in the Newt (18), or partly epiblast and partly differentiated yolk-cells in the Anura (1, p. 102), whilst the lateral walls and the floor of the archenteron are formed by modified yolk-cells.

Houssay (26) and Moquin-Tandon (38) are, so far as we are aware, the only authors who have combated this supposititious mode of formation of the archenteron. Houssay examined the Axolotl and Moquin-Tandon several Anura. They both state that the archenteron is formed by splitting amidst the yolk-cells, from which it follows that both in the Urodela and Anura there is no invagination of epiblast into the yolk, and that the archenteron is surrounded by modified yolk-cells which eventually form a distinct hypoblastic layer, which is continuous with the epiblast at the margin of the blastopore.

The blastopore is unanimously defined in Amphibians as an opening, round the margins of which the epiblast and hypoblast are continuous and which forms the passage of communication between the archen teron and the exterior. But concerning the formation and surround ings of the archenteron there are two opposed opinions, which are—

1. That the archenteron of the Amphibia is a cavity, formed as in Amphioxus by invagination, and is lined partly by modified yolk-cells and partly by invaginated epiblast.

2. That the archenteron is formed *in situ* by splitting amongst the yolk-cells, and that it is entirely surrounded by modified yolk-cells.

To these opinions it will be necessary to return, but in the meantime we pass to a consideration of the various accounts of the fate of the Amphibian blastopore.

This opening, after remaining for a time, is, according to Balfour's account, decreased in size by the approximation of its lips until it forms "a narrow passage, on the dorsal side of which the neural tube opens. . . . . The external opening of this passage gradually becomes obliterated, and the passage is left as a narrow diverticulum leading

from the hind end of the mesenteron into the neural canal (1, p. 108).
It forms the post-anal gut, and gradually narrows and finally atrophies."
This account is in the main conformable with Goette's first description
of Bombinator: "Die verengte sich vorherrschend von beiden Seiten
her, so dass sie spaltartig wurde und ihr Längsdurchmesser in der
Medianebene des sich entwickelnden Embryonalkörper lag " (14, p.
132). And, further, Balfour (1, p. 108) states that the anus is
formed in *Rana temporaria* at an earlier period than in Bombinator, as
described by Goette, but in the same manner, that is, by the fusion of
a diverticulum from the mesenteron with a cutaneous invagination.
Almost the same change takes place in *Rana fusca* (46); Spencer,
however, as a result of his study of *Rana temporaria*, came to the
conclusion that the blastopore was transformed into the permanent
anus (52); and Alice Johnson and Lilian Sheldon (28) arrived at a
similar conclusion concerning the blastopore of *Triton cristatus*, which
according to the account of Scott and Osborn (48), is enclosed by the
neural folds.

Durham (9) describes a neurenteric canal and a blastopore present
at the same time in Rana; and Sidebotham (51) asserts that the
neural folds do not enclose the blastopore, which is closed subsequently
to the meeting of the neural folds, and that the anus is derived from
an independent proctodaeal invagination.

Morgan's (39) interesting observations tend to a solution of the
difficulty caused by the preceding contradictions, inasmuch as they
show that in *Amblystoma punctatum* the mesial portions of the lateral
lips of the blastopore are first approximated, the blastopore thus
becoming hour-glass-shaped, and then fused, so that the previously
single opening is divided into two; the anterior of the two is enclosed
by the medullary folds, and becomes the neurenteric canal; the
posterior remains as the anal orifice. Goette (15) has lately described
a somewhat similar condition in Bombinator.

In *Rana halecina*, according to Morgan (39), there is behind the
blastopore a groove terminating posteriorly in a dark spot, which is a
later formation than the blastopore. In describing a very similar
groove in *Bufo lentiginosus* he states that it is separated from the
archenteron by three embryonic layers, a point of importance in any
comparison of this region with the primitive streak of other Vertebrates.

Erlanger's (10A) observations upon the Anura have led him to the conclusion that the blastopore closes both from before backwards and from behind forwards. The closure from before backwards is associated with the formation of the primitive streak, which lies in front of the neurenteric canal. The closure from behind forwards gives rise, in the first instance, to a fused mass of cells, which rapidly differentiates into layers, after which the anus forms in the situation previously occupied by the posterior portion of the blastopore. Therefore, so far only as the statements of the authors to whom we have referred are concerned, the blastopore of Amphibians may either—

Become (1) gradually constricted, being transformed into a narrow canal, which for a time unites neural and alimentary cavities, but eventually disappears—Balfour (1), Schultze (46). Scott and Osborn (48);

Or (2) it is transformed into the anus—Alice Johnson (27), Spencer (52);

Or (3) it is not enclosed in the neural folds, it does not form the anus, but gradually disappears—Sidebotham (51);

Or (4) the anterior part becomes the neurenteric canal and the posterior part the anus—Morgan (39), Schwarz (47), Goette (15);

Or (5) the anterior portion becomes the primitive streak, the middle portion the neurenteric canal, and the posterior part, after being closed, is again reopened in a small portion of its extent as the anus—Erlanger (10A).

But this summary does not include all the fates which are allotted to the blastoporic opening, for its history has been intimately connected with that of the primitive streak and the axial line of the embryo by the researches of His (22 and 23) and Rauber (42), whose observations resulted in the "concrescence theory" of Vertebrate development, which Minot (36) has striven to support in his recent papers. It is necessary to consider this theory, as the results of our observations bear upon it; but before referring further to it we must draw attention for a moment to the structure of the lips of the blastopore. It is almost universally allowed that in all Vertebrates, with the exception of Amphioxus, there is in this region a fusion of all three layers of the germ, and it follows as a consequence, if the lateral borders of the blastopore are approximated and united, that there should result a mesial axial rod of fused tissue, which cannot in the first instance be resolved into distinct layers. Such a fusion of the blastopore lip was long ago shown to occur in Elasmobranchii by Balfour (1, p. 51), who compared the line of fusion of the lips of the Elasmobranch blastopore to that portion of the blastodermic areas of the Avian ovum to which the term primitive streak had been applied, "The primitive streak represents the linear streak connecting the Elasmobranch embryo with the edge of the blastoderm after it has become removed from its previous peripheral position, as well as the true neurenteric part of the Elasmobranch blastopore" (1. p. 238). . . . "That it is in later stages not continued to the edge of the blastoderm, as in Elasmobranchii, is due to its being a rudimentary organ" (1, p. 240). In Balfour's opinion, therefore, the primitive streak represents a portion of the blastoporic opening situated posterior to the neurentric canal, and in conformity with his statement that the anus of Rusconi in the frog is the whole of the blastopore, which becomes gradually contracted, he does not describe a primitive streak in Amphibia. His conclusions, however, are not in accord with those of the upholders of the concrescence theory, who believe that the axial portion of the embryo is formed by the fusion of the lips of an elongated blastopore. The fusion takes place from before backwards, and results in the formation of an axial area, which lies in front of the remains of the blastopore. This area is termed the primitive streak.

If we examine the anus of Rusconi of Amphibia in the light of Balfour's conclusions concerning the blastopore, and then in that of the concrescence theory, we are forced to believe in the first case that either there is no primitive streak in Amphibians, or that it lies somewhere posterior to the neurenteric canal; and, further, that it would be completely represented by the fusion of the middle part of the lateral lips of the blastopore in *Amblystoma punctatum* (39), and in Bombinator according to Goette's description; or, in the second case, that the anus of Rusconi represents the posterior part of the blastopore, and that the primitive streak is situated in front of it.

Schwarz (47) and Goette (15) compare the "Prostomaschluss" which lies between the neurenteric canal and anus in Bombinator to the primitive streak of the higher Vertebrates.

Morgan (39), Alice Johnson (27), and Schultze (15) mention a primitive streak in front of the blastopore. Morgan gives no description either of the formation or structure of the streak. Alice Johnson (27) describes a fusion of the germinal layers in the region in front of the blastopore which she calls the primitive streak, but she does not say whether the fusion is primary or secondary. There is, however, no doubt about the formation of the area called primitive streak by Schultze in his description of Rana Fusca, for he describes the process in the following words :—" Während auf den vorhergehender Entwicklungsstadien dorsal und median das äussere Keimblatt durch einen Spaltraum von dem mittleren Blatt getrennt war, und beide Blätter nur an der ringformigen Zone der Blastoporuslippe in einander übergingen, bildet sich gegen Ende der Gastrulation, von der Mitte der dorsalen Urmundlippe aus, eine nach dem Kopfe hin vorwärts schreitende lineare Verwachsung, des ausseren und mittleren Blattes aus". . . . . "diese lineare Verwachsung, in welcher wie in dem Primitivstreifen des höheren Wirbelthiere Ektoblast und Mesoblast zusammenhängen, von der dorsalen Urmundlippen sich nach dem Kopfe hin allmählich ausdehnt, und wächst also der Primitivstreifen auch bei den Amphibien von hinten nach vorn " (45, p. 330).

This statement of Schultze's tends to increase the difficulty of the situation, for it agrees neither with Balfour's nor with the concrescence theory of the primitive streak. In opposition to Balfour's theory Schultze places the primitive streak in front of the neurenteric canal, and he says that the streak is formed from behind forwards, a statement incompatible with the concrescence theory, which necessitates that " the entodermal canal and primitive streak begin at the edge of the blastoderm and grow at their posterior end away from the segmentation cavity, and at the same rate the blastoderm overspreads the yolk " (36, p. 511).

Neither do Erlanger's observations (10A) tend to simplify the problem. He associates the primitive streak with the antero-posterior closure of the dorsal portion of the blastopore, but according to his figures the primitive streak ultimately exceeds in length the diameter of the primitive blastopore. He figures no sections to show the structure of the streak, and we are inclined to believe that he has mistaken the neural furrow for the primitive streak. But if his conclusions are correct, they support the concrescence theory only in part, in so far as the fusion from before backwards of a portion of the blastopore is concerned; whilst the fusion of the lip of the ventral portion of the blastopore, which he describes as taking place from behind forward, is at variance with the concrescence theory.

Nevertheless, to the summary which has already been given of the fate of the blastopore in Amphibians (p. 457), we must add the further possibility that, according to the concrescence theory, the posterior part of it only is to be recognised in the anus of Rusconi, the lateral lips of the anterior part having undergone a "retrogressive fusion" similar to that described by His in bony fishes (22), and that this fusion is continued until the complete obliteration of the blastopore results.

From the statements noted in the preceding survey of the literature of the subject we may deduce the following summary of the presence or absence and the mode of formation of the primitive streak in Amphibia.

1. There is no mention of a true primitive streak—Balfour (1).

2. There is a primitive streak situated in front of the anus of Rusconi—Schultze (45), Minot (36), Alice Johnson (27), Erlanger (10A).

3. The fused lips of the blastopore between the neurenteric canal and the anus represent the primitive streak—Schwarz (47).

The primitive streak is formed—

1. By retrogressive fusion of the lips of the blastopore (concrescence theory).

2. By fusion of the lips of the blastopore between the neurenteric canal and anus—Schwarz (47).

3. By fusion of the epiblast and mesoblast from behind forwards in front of the anus of Rusconi—Schultze (45).

Before concluding this short survey of previous records we wish to summarise the statements, some of which have been already referred to concerning the formation of the anus in Amphibians.

1. It is a new formation in the region below the blastopore—Balfour (1), Sidebotham (51).

2. It is the blastopore—Sedgwick (49), Alice Johnson (27), Spencer (52).

3. It is the posterior portion of the blastopore—Morgan (*Amblystoma punctatum*), (39), Schwarz (47), Goette (15).

4. It is a secondary opening in the situation of the posterior part of the primitive blastopore—Erlanger (10A).

The Enclosure of the Yolk by Epiblast and the Formation of the Archenteron.

At the period generally spoken of as the end of the segmentation the anural ovum is a sphere which contains an excentrically situated segmentation cavity.

The segmentation cavity has a roof which ultimately becomes the anterior wall of the gastrula; for the anus, which marks the posterior end of the embryo, appears at the opposite pole of the ovum, that is in the floor of the segmentation cavity.

The roof of the cavity is formed by two or three rows of comparatively small pigment-bearing cells, and the floor by a mass of ill-defined nutriment-laden cells which collectively form the yolk. At the margin of the segmentation cavity the two walls merge into each other by a series of insensible gradations, so that it is impossible to say where one ends and the other begins. Nevertheless it is certain that the cells of the roof of the segmentation cavity are epiblast, for they ultimately form a portion of the external covering of the embryo; but it is incorrect to speak of the opposite wall, the yolk-cells, as modified hypoblast (1, p. 103), unless it is allowed that epiblast, may be formed from modified hypoblastic cells. For during the formation of the blastopore the epiblast does not grow over the yolk-cells, enclosing them by a process of epibolic invagination. If it did, it would be possible to recognise on section a line along which the epiblast terminated. No one has described or figured such a line of limitation except diagrammatically; and a careful examination of thin sections of anural ova, in the stages preceding the completion of the circular blastopore, shows clearly that the descriptions given of the gradual extension of the epiblast over the yolk represent theory rather than fact.

The changes observable are, firstly, that the cells of the superficial layer of the yolk gradually become more distinctly pigmented; and, secondly, that they thereafter divide into smaller and more distinct segments, which arrange themselves into two layers, a superficial layer of somewhat cubical deeply pigmented cells, and a deeper layer of less pigmented and more rounded cells, irregularly arranged into two or more rows: the latter are separated by a distinct space from the subjacent remainder of the yolk-cells. This transformation and re-arrangement proceeds backwards towards the blastopore, stopping about ·352 mm. from that aperture, first on its dorsal and then on its lateral and ventral borders, where an area of fusion remains, which is for convenience divided into a dorsal, a ventral, and two lateral lips or borders.

We conclude, therefore—(1) That the segmentation of the anural ovum does not result in the formation of a vesicle, the roof of which is epiblast and the floor modified hypoblast, but that at the end of the segmentation the primary layers of the ovum are only partially formed. The roof of the segmentation cavity is epiblast, but the floor or yolk is not modified hypoblast. It consists of undifferentiated germ-cells, the true characters of which are not at first recognisable.

(2) That the yolk is not enclosed by the gradual extension over it of a previously differentiated epiblast, but that the superficial layer of yolk-cells becomes gradually differentiated into the two layers of epiblast, leaving a remainder of the yolk. This consists of hypoblast and mesoblast, which are not at first separated from each other.

The formation of epiblast from yolk-cells does not take place over the whole surface of the yolk, for at the posterior pole of the ovum there is a circular patch of yolk-cells from which no epiblast is formed. The cells in question form the yolk-plug, and when the epiblast formation has arrived from all sides at their margin, the circular blastopore or anus of Rusconi is first definitely established.

During the period of differentiation of the epiblast the archenteron has also been forming. The first indication of this cavity is the appearance of a curved area of pigmentation (fig. 1 *a* and fig. 2 *AR*) of semilunar outline amidst the yolk-cells at the posterior pole of the ovum above the equator. The convexity of the pigmented area is directed forwards amidst the yolk-cells, towards the segmentation cavity. The angles are turned ventro-laterally. The centre of the concavity of the semilunar area corresponds to the dorsal lip of the blastopore. As the convexity of the area extends forwards towards the segmentation cavity, its lateral angles travel ventrally along the lines of the lateral lip of the blastopore until they meet in the situation of the ventral lip of that opening, which is thus marked out by a line of pigmentation that indicates the situation of the future cavity.

The pigmented area is produced and extended by the deposit of pigment in the adjacent margins of a double row of yolk-cells in the manner shown in figs. 1 and 2, which are drawings of sections of the advancing anterior extremity of the pigmented area of a much more developed ovum   The pigment is deposited in the protoplasm of the double row of cells which, eventually, will form the boundary wall of the archenteron, and it also radiates from this area along the adjacent margins of the cells of each row.

It is to be understood, however, that directly after the pigmented area is first formed, and as it extends forwards and ventrally, a slit-like space appears in the middle of its posterior portion.   This space first limits the dorsal lip of the blastopore, and then extends forwards and ventrally, following the deposit of pigment, and separating the two rows of marginally pigmented cells from each other.   It is the archenteric space.   As its lateral angles extend ventrally along the line of pigmentation they define the lateral lip of the blastopore; their union ventrally gives rise to the ventral lip, and at the same time defines the posterior limit of the ventral wall of the archenteron (fig 3).

A diagrammatic representation of a sagittal mesial section of the ovum at the period of completion of the blastoporic lip is given in fig. 4, from which it will be readily seen that, at this period, the ventral wall of the archenteron extends from the point a to the point x. In a portion of its extent the floor of the slit-like cavity lies parallel to the superficial surface of the ovum, but posteriorly it is projected outwards into a deficiency in the external wall (the blastopore), forming in this region the yolk-plug or bouchon d'Ecker.   At the same stage the dorsal wall of the archenteron is not co-extensive with the ventral.   It extends only from the point a to the point b, which marks the dorsal lip of the blastopore.   From the dorsal to the ventral lip of the blastopore the wall of the archenteron is deficient.   It will be shown afterwards how this portion of the wall, which we shall call the posterior wall of the archenteron, is completed.

At present we more particularly desire to call attention to the fact that the point x, which at this period marks the posterior limit of the ventral wall of the archenteron, remains fixed throughout the later periods, that eventually it is situated just in front of the anal orifice, and that there is no extension ventrally of the archenteron beyond it, such as that described by Balfour, who, speaking of the gradually narrowing blastopore, says : " At its front border, on the ventral side, there may be seen a slight ventrally directed diverticulum of the alimentary tract, which first becomes visible at a somewhat earlier stage " (1, p. 108). On the contrary, the ventral wall of the archenteron is only increased in front of the ventral lip of the blastopore by the forward extension of the anterior end of the cavity and the growth of the ovum, and is completed immediately in front of the ventral lip, so far as we have been able to ascertain, by the gradual withdrawal of the yolk-plug and the rearrangement and differentiation of its constituent cells.

It is certainly noteworthy that the slit-like archenteron does not appear in the midst of the yolk-cells until the epiblast reaches the region of the blastoporic margin ; and it is this fact, together with a predisposition to discover, if possible, an invaginative process, which seems to have led to the description of the formation of the archenteron of the Anura by invagination of the epiblast.

A careful examination of the walls of the extending archenteric cavity reveals no evidence in support of this ideal invaginative process ; on the contrary, it shows that even at the period of completion of the blastoporic opening (see figs. 1 and 3) the greater part of the cavity is surrounded by large nutriment-laden, marginally pigmented cells. More especially is this the case in the anterior portion of the, as yet, incomplete cavity (fig. 2), and at its posterior extremity (fig. 3), where the slit-like space is only just formed. In the latter situation the only cell that can be considered as distinctly epiblastic is that marked c. The remaining cells are undoubtedly large yolk-cells ; and it is only by the division of these cells, and the arrangement of some or all of their descendants into a distinct layer, that the true hypoblastic lining of the alimentary canal and its diverticula is formed. As the completion of the definite hypoblast is intimately associated with the separation of the mesoblast and chorda, its further history cannot at this period be entered upon, and we may therefore proceed to the termination of this section by a summary of the conclusions deduced from that which has been here set forth.

1. The blastopore is a deficiency in the posterior wall of the archenteron.

2. The archenteron of the Anura is not formed by invagination, but by a process of splitting amongst the yolk-cells very similar to that described by Houssay (20) in the Axolotl.

3. The situation of the archenteric cavity is first defined by the desposition of pigment in the adjacent margins of a double row of yolk cells.

4. No portion of the archenteric wall is formed by invaginated epiblast; on the contrary, the archenteron is surrounded in the first phases of its development by large yolk-cells, which eventually give rise to the definite hypoblast.

5. The ventral lip of the blastopore indicates the posterior end of the primitive ventral wall of the archenteron.

6. There is no ventral and forward extension of the archenteron in front of and below the ventral lip of the blastopore by the production of a diverticulum in that situation.

7. The ventral wall of the archenteron, in front of the ventral lip of the blastopore, is completed by the extension of the anterior end of the cavity, and by the withdrawal and modification of the cells of the yolk-plug.

Manner in which the Anus of Rusconi closes.

Having discussed the changes that take place in the ovum up to the time of the formation of the large circular blastopore or anus of Rusconi, we will now proceed to describe the manner in which, according to our observations, the closure of the blastopore or anus of Rusconi seems to be effected, and to give in some detail an account of the structures which are the immediate result of the method of closure about to be described.

According to some of the former accounts, to which we have made reference above, the anus of Rusconi has been said to diminish in size by the gradual coming together of each portion of the blastoporic rim simultaneously. This we believe to be incorrect. No doubt from surface views alone the hitherto accepted accounts receive much support.

We have, however, paid particular attention to this point, and have come to the conclusion that the anus of Rusconi gradually diminishes in size by the concrescence of the ventral part of the lateral lips, as indicated in figs. 10 and 18. In these diagrams, the space lettered *BR*, inclosed between two concentric rings, represents the blastoporic rim at the time of the completion of the circular blastopore, or anus of Rusconi; and the shaded space, lettered *BR'*, represents in fig. 10 the position of the blastoporic rim soon after the anus of Rusconi has begun to diminish in size, and in fig. 18 a later stage, when the anus of Rusconi has "contracted" so much that its diameter is now only about one third of its original size.

Figs. 7, 8, 9 are from sections taken nearly horizontally through an embryo in which there was no trace of neural folds, and in which the anus of Rusconi had only slightly contracted.

Fig. 10 is a diagram of the same embryo representing the changes which we conclude must have taken place.

The blastoporic rim, round which the three layers, epiblast, mesoblast, and hypoblast, are all fused, which occupied, on the completion of the circular anus of Rusconi, the position indicated by the space *BR* between the two concentric circles, had in the case of the embryo examined come to occupy the position of the shaded portion lettered *BR'* in the diagram. In the same diagram the line *x* represents that portion of the original anus of Rusconi which has become closed up by the concrescence of the ventral part of the lateral portions of the rim, as stated on p. 466.

We were unable in this specimen to find any trace of the line of fusion *x* on the surface, and as yet there was no trace of the neural folds. Accordingly, in order to be able to cut our sections as near as possible in the horizontal plane as desired, we had to dissect away the opposite pole of the embryo, and then by observing the shape of the yolk-plug we were able to cut our sections very nearly as required.

Fig. 7 has been drawn from a section (which is not quite horizontal) taken through the centre of that portion of the blastopore which was still open. At the lips of the blastopore the three layers are seen to be fused.

The line running from *ZZ'* ends at the furthest point from the blastoporic lip to which the fusion of epiblast (*EP*) and mesoblast (*ME*) extends on one side, and the line from ZZ ends at the furthest point from the blastoporic lip to which the fusion of those layers extends on the other side. It will be noticed that on one side the fusion extends further than on the other. No doubt this is chiefly owing to the section being not quite horizontal, the side of the ZZ being the more ventral of the two. We have, however, taken the thicker side for the purpose of measurement.

If the diminution in size of the blastopore has been produced by the concrescence of the ventral portion of the lateral lips as suggested above and as diagrammatically shown in figure 10, it is clear that the fusion of layers ought to extend further from the actual lip of the blastopore on the ventral border of the blastopore than on the sides or dorsal border. This will be clearly seen by reference to the diagram, fig. 10.

It is equally clear that this would not be so if the closure of the blastopore were strictly centripetal, in which case the fusion of layers should extend on all sides equally.

If the distance be measured across the blastopore in fig. 7 with a pair of compasses, it will be found to be about the same distance as between the edge of the blastopore lip and the point of furthest extension of fused epiblast and mesoblast on the side of ZZ.

In the series of sections from which this fig. 7 and the two represented in figs. 8 and 9 were selected, the open blastopore was present in forty sections.

Assuming the open blastopore to have been circular, we may conclude that it would have required forty sections to cut through the fused mass between the edge of the blastoporic lip (B) and the furthest extension of the fused mass indicated by the line drawn from ZZ, for we have just now seen that the distance from B to B is about the same as from B to ZZ. In other words, the forty-first section ought to show no, or but little, fusion of layers.

Fig. 9 has been drawn from the forty-first section, and instead of there being no fusion of layers there is an extent of fused layers about equal to the two fused blastoporic rims, or indeed, a little more.

At the sixtieth section the epiblast and mesoblast are still distinctly fused, and it is not until the eightieth section below the edge of the ventral lip of the blastopore that the layers can be definitely said to be separate. In short, whereas by calculation the fusion of layers on the

G

lateral lips will be cut through in forty sections, it requires close upon eighty sections to cut through the fused mass at the ventral lip.

In fig. 10 the levels of the sections (figs. 7, 8, 9) are indicated by the lines 7, 8, 9.

Thus a sagittal section through the blastopore ought to show a very much larger mass of fused layers at its ventral lip than at its dorsal.

This is seen to be the case as shown in fig. 5, which is a sagittal section through the blastopore of a rather older frog embryo. The blastopore has closed to a considerably greater extent than was the case in the embryo we have just been describing. The actual extent of the closure of the blastopore by the concrescence of the latero-ventral lips is represented, we think, by the distance between the point x and the present edge of the ventral lip of the blastopore.

We find by a series of measurements that at whatever stage during the closure of the blastopore the section be taken, the distance between point x and the edge of the dorsal lip of the blastopore is always approximately the same, and the same as the diameter of the blastopore at its first commencement. We say approximately, for, owing probably to variation in size of the egg, the measurements do not exactly agree.

Fig. 11 was drawn from a living specimen in which the blastopore had become greatly reduced, and in which the neural plate and neural groove were distinctly visible along the dorsal surface.

From the lower, or ventral, lip of the open blastopore stretches a very faint (too deeply drawn in the figure) line, which probably is the remaining trace of the line of concrescence of latero-ventral blastoporic lips.

Fig. 13 is a section taken at right angles to this line, not of the same embryo, but of one about the same age. In this the line is seen to be a groove (PG) on the surface, below which all these layers are fused.

We will at once call this groove *primitive groove*, and the fused mass below it *primitive streak*; for, as we hope to prove in the sequel, there can be no reasonable doubt that these structures are the homologue of those parts in the chick embryo to which these names were originally given.

A section taken *dorsal* to the blastopore, and dorsal to the fusion of ayers at the dorsal lip of the blastopore, is seen in fig. 12.

Here the epiblast is seen to be quite distinct from the mesoblast. There is a distinct groove, the neural groove, very different in character from that of the groove ventral to the blastopore, or primitive groove, while the epiblast is thickened on either side to form the neural plate. The level of the respective sections is marked in fig. 11 by the lines numbered 12 and 13.

Before discussing the relations between the structures, the formation of which we have been following in the last two or three pages, and which we have named primitive streak and primitive groove, and the primitive streak and primitive groove of the chick, we will follow its history a little further, until the stage at which we may say it has attained its greatest development.

We must, therefore, describe the condition of the primitive streak of a frog embryo of about $2\frac{1}{2}$ mm. in length; that is to say, an embryo in which the neural folds are on the point of meeting, or have just met, as shown in fig. 23.

Primitive Streak of the Frog at the Time of the Closure of the Neural Folds.

Fig. 23 is the surface view of the posterior end of a $2\frac{1}{2}$mm. frog embryo, in which the neural folds have just met.

It was drawn partly from a preserved specimen, and partly from a model, as described on page 83.

The line of junction of neural folds (J) is marked by a deep groove ending posteriorly in the blastopore.

The blastoporic opening, as seen from the surface, is no longer circular, but is more or less lozenge-shaped, due to the folding its dorsal lips have undergone (as will be described later).

The lengthening of the embryo has removed from the blastopore all trace of the "yolk-plug."

Running ventrally from the blastopore is the primitive groove, a rather narrow but now sharply defined groove near the ventral end, which is suddenly deepened into a pit.

Beyond the pit the groove continues a short distance. This pit is the commencement of the anus, which will shortly perforate at this spot.

In fact, in the model from which this figure was partly drawn the anus had just perforated. In the sections, however, about to be described as representing this stage the anus has not quite perforated. On either side of the primitive groove a distinct ridge is visible.

The Bracket (*PS*) in the figure includes the whole of the primitive streak from end to end.

The sections 14, 15, 16, 17, are horizontal sections taken along the lines 14, 15, 16, 17, in fig. 23, and are, therefore, taken at right angles to the longitudinal axis of the primitive streak.

Fig. 14 is a camera drawing of a section through the blastopore (line 14, fig. 23), which is nearly closed in.

In this section the epiblast, mesoblast, and hypoblast are seen to be all fused at the lips of the blastopore. It may be noted that the mesoblast, close to the lip of the blastopore, seems to be divided into two layers: a very dense compact layer (*ME*) next the epiblast, in which the cells are so closely packed as to render it quite impossible to draw each cell by camera; while inwardly the mesoblast-cells (*ME"*) are much looser, and are more deeply pigmented and easy traceable by help of the camera. In fact, one is almost inclined to say that the former are mesoblast-cells of epiblastic origin, while the latter are mesoblastic cells of hypoblastic origin.

Fig. 15 is a section taken out about midway between the blastopore and the anal pit; that is to say, across the middle of the primitive streak. This section is directly comparable to that just described, *except that the lips of the blastopore have met and fused*, as described in the earlier part of this paper. The surface is distinctly grooved, the edges of the groove being raised slightly into ridges. The same feature in the mesoblast may be noticed here as in fig. 14.

Fig. 16 is taken through the anal pit.

Here the primitive groove is very much deepened, so that not more than two or three cells prevent the completion of the anal perforation. This pit is deeply pigmented, and heavy deposits of pigment line the few intervening cells between exterior and archenteron. All three layers are still fused.

Fig. 17 represents a section taken below the anal pit; that is to say, it was the thirteenth section after the last one figured, fig. 16. In this, although behind the anus, there is undoubted fusion of epiblast and mesoblast. It is altogether ventral to the archenteron, so that there can be no fusion with true hypoblast. There is no fusion either with the yolk-cells. *The surface is grooved, as in the sections anterior to the anus.*

The fusion of layers continues through six more sections; that is to say, there are about twenty sections ventral to the anus, in which there is fusion of epiblast and mesoblast; in other words, *the anus is a perforation through the primitive streak.*

Comparison of the "Primitive Streak" of the Frog with the Primitive Streak of the Chick.

We have now described the history of the blastopore or anus of Rusconi from the time of its first formation until it has been reduced to a comparatively minute passage between the archenteron and exterior. We have described this change as being due to the concrescence of the lower lips of the blastopore, as shown diagrammatically in figs. 10, 18, 19, whereby there is produced a median streak characterised by the fusion of layers, along the surface of which lies a groove, stretching from the ventral lip of the remaining blastopore ventralwards as far as the original extension of the blastopore or anus of Rusconi. To this structure and to the groove upon it we have applied the terms primitive streak and primitive groove respectively.

If one of our sections figured, *e.g.* fig. 15, is compared with the figure of a transverse section of the primitive streak of a chick on p. 155 of Balfour's 'Comparative Embryology,' vol. ii (second edition), the resemblance between the two sections is very marked.

In each case there is an intimate fusion between epiblast and mesoblast, and a more uncertain fusion between mesoblast and hypoblast. In each case the surface is marked by a groove.

Nor is it at all impossible that the loose pigmented cells noticed above, and lettered *ME* in figs. 14 and 15, may be compared with Balfour's "layer of stellate cells" shown in the figure in the 'Comparative Embryology' referred to. In both cases they seem to arise from the hypoblast rather than the epiblast. In the frog, however, they are unrecognisable a short distance from the streak.

In comparing the two streaks with regard to their relations to the rest of the embryo, we find that we must use the term primitive streak in a rather wider sense than it is usually used.   In the chick the anterior limit of the primitive streak may be said to be marked by the posterior end of the notochord with which the streak is fused.

The structure we have so far called primitive streak runs from the *ventral lip of the blastopore.*   It is, however, obvious that had the concrescence continued a little further, so that the *whole* anus of Rusconi had been obliterated by the fusion of the blastoporic lips, the primitive streak would have then commenced at the posterior end of the notochord, as in the chick ;   for the notochord and neural folds in the frog are continuous with—that is to say, fused with—the dorsal lip of the blastopore.   It is, therefore, clear that *the homologue of the primitive streak of the chick is in the frog the whole of the blastoporic lip, whether fused or not.*

Relation of Anus to Blastopore.

The relation of the anus to the blastopore has been a much controverted subject.   Goette (14) and Balfour (1) described it as an entirely separate perforation of the body-wall ventral to the blastopore ; Spencer (52) stated that the blastopore became directly the anus ; Sidebotham (51), who gave a most careful and exact account of the facts, agreed with Goette and Balfour on this point, and, failing to recognise the primitive streak in the frog, described it as they had done, as a new opening independent of the blastopore.   In Erlanger's (10a) opinion it is a secondary opening in the situation of the posterior part of the original blastopore.   Those who have followed our account so far will observe that it agrees most closely with those of Goette, Balfour, and Sidebotham on this point ;   but, *since the perforation takes place within the primitive streak,* the conclusions we draw from the facts are different.   We infer, therefore, with Erlanger, that the anus of the frog, although apparently a new perforation, is really a re-opening of a temporarily closed portion of the original blastopore.

Since the perforation occurs at the base of the diverticulum of the archenteron (which, it will be remembered, was formed by the closing in of the ventral portions of the lateral lips of the blastopore), it may be supposed that *it is the most ventral end of the blastopore which, morphologically speaking, persists as the anus.*

The Primitive Streak of the Frog and other Vertebrates.

The term primitive streak appears to have been first applied to the dark line which appears in the avian blastoderm at the posterior part of the area pellucida.   This line is generally looked upon as the optical expression of a linear thickening and fusion of the blastodermic layers, though Kölliker (31, p. 134) maintains that it is due to proliferation of the epiblast-cells alone.   It seems certain, however, that it is impossible, during certain periods of the growth of the embryo, to distinguish the three germinal layers from each other in the area of the primitive streak (Duval, 10, p. 181).

After a time differentiation proceeds in the streak ; a groove appears on its superficial surface, and this is deepened anteriorly into a perforation, which ultimately becomes the neurenteric canal (47). The anterior wall of this perforation is formed by a mass of cells in which the epiblast and hypoblast are united (12).   In the posterior portion of the streak the anal membrane is developed by the gradual thickening and apposition of the hypoblast and epiblast (13, pl. x, figs. 4 and 5, p. 299 ; and 47, pl. xiv, fig. 81, p. 203).

The lateral margins and posterior extremity of the streak are continuous with the mesoblast, which appears to grow out from them into the surrounding area.

The anterior end of the streak is continuous with the chorda ventrally, and the central part of the neural plate dorsally.   In the region surrounding the primitive streak the three layers of the blastoderm are distinct from each other, except in front of the anterior extremity of the streak, where the so-called "Kopffortsatz," the first rudiment of the chorda, is still continuous with the entoderm. The connection between the hypoblast and the mesoblast, which exists throughout the whole length of the streak, is first dissolved posteriorly, where for a certain period the epiblast and mesoblast are fused, but the hypoblast forms a distinct layer.

Between the primitive streak of a bird and the frog there are resemblances and differences of importance.   In the frog the primitive streak is formed by a concrescence of the lips of the blastopore, which proceeds from behind forwards, and which is only completed on the obliteration of the neurenteric canal.   In the bird the primitive streak is formed from before backwards according to Duval (10) and Schwarz

(17), from behind forwards according to Balfour and Deighton (2) and Koller (30); and the appearance is due apparently to thickening and fusion of the two primary layers—Duval (10), Balfour and Deighton (2).

In both the frog and the bird the lateral margins and posterior extremity of the streak are continuous with the mesoblast, which lies free between epiblast and hypoblast outside the area of the streak.

In the frog the anterior wall of the neurenteric canal is bounded by an area of fusion, in which the middle of the neural plate and the posterior end of the chorda are united. After the obliteration of the neurenteric canal the posterior end of the chorda and the centre of the neural plate are continuous with the anterior end of the primitive streak.

In the bird the anterior end of the primitive streak is at first continuous with the centre of the neural plate and the " Kopffortsatz," and after the formation of the neurenteric canal the chorda and the neural plate are fused in the anterior wall of the latter orifice, becoming again continuous with the anterior end of the primitive streak after the disappearance of the neurenteric canal.

In the frog the anus is formed in the posterior part of the primitive streak. It is a reopening of a portion of the closed blastoporic orifice. It is not the remains of the blastopore, as in *Amblystoma punctatum* (39) and Bombinator (15). The anus of the bird is morphologically equivalent to the anus of the frog; and it is also formed, in all probability, by a reopening of a previously closed orifice (10).

In reptiles, according a Kupffer (33), there is no primitive streak, but in the posterior part of the embryonic area the epiblast is invaginated, and the mesoblast arises, in part at least, from the margins of the invagination. The cavity of invagination is the archenteron, and the superficial opening the Urmund, which would thus entirely correspond to the anus of Rusconi in Amphibians.

But Balfour (1, p. 168), Strahl (54), Hoffmann (24), Weldon (56), Mitsukuri and Ishikawa (37), and Wenckebach (57) describe a primitive streak. Balfour states that in the primitive streak of lizards the epiblast and hypoblast are fused, though the greater part of the streak consists of proliferated epiblast. Wenckebach, however, looks upon the hypoblast as a purely passive agent in the formation of the primitive streak in lizards; and in the tortoise, according to Mitsukuri and Ishikawa, the streak consists, in the first instance, mainly of a mass of hypoblast or yolk, which they compare to the yolk-plug of Amphibians. To this, however, it cannot correspond, for we have already shown that the yolk-plug of Rana is a portion of the ventral wall of the archenteron, whilst the primitive streak is formed by the fusion of the lateral lips of a deficiency in the posterior wall of the same cavity.

But, whatever the mode of its formation may be, the streak eventually becomes perforated, both anteriorly and posteriorly: anteriorly by the neurenteric canal—Balfour, Hoffmann, Weldon, and Strahl;[1] and posteriorly by the anus—Weldon. It is thus, in all essential respects, comparable with the primitive streak of birds.

In mammals also a primitive streak is found. It is described as commencing in the sheep (5) and in the shrew (25) as a knob-like swelling of the epiblast in the posterior part of the embryonic area. The epiblastic thickening extends backwards and terminates in a tail-swelling. According to Bonnet (5), Hubrect (25), and Hensen (20), the two primary layers fuse in the streak, though it does not seem certain that the hypoblast takes any part in the thickening. Kölliker (31), denies that the hypoblast takes part in the formation of the streak in the rabbit. Rabl (41) agrees with Kölliker, and Fleischmann (11) makes a similar statement concerning the cat. In the primitive streak of the guinea-pig (29) there is fusion of the layers anteriorly, but posteriorly the thickened epiblastic ridge is separate from the hypoblast. In the mole (17) all the layers take part in the formation of the streak, but after a time they are fused only at its anterior and posterior ends, whilst in the middle the hypoblast forms a distinct layer. At the anterior end of the mammalian primitive streak more or less distinct traces of a neurenteric canal have been found in the rabbit by Strahl (55), in the mole by Lieberkühn (35) and Heape (17), in the sheep by Bonnet (5 and 6), and in the bat by Van Beneden (3). In the anterior wall of the opening the chorda and neural epiblast are fused, and laterally and posteriorly the mesoblast hangs in connection with the margins of the streak.

[1] Strahl's statement (54), that in L. agilis the perforation occurs in the middle of the streak, simply means, apparently, that there is a region of fusion between the epiblast and hypoblast in front of the opening.

The anal membrane is formed in front of the posterior end of the streak in the rabbit—Strahl (55); in the mole—Heape (17); in the guinea-pig—Keibel (29); and in the rat and mouse—Robinson (43). In the sheep, Bonnet (7) states that the anal membrane is situated at the posterior end of the streak ; and Rabl (41) figures it in the same position in the rabbit.

The differences described are slight and probably unimportant, and the main facts stand out clearly. As in the Sauropsida, the primitive streak is a line of fusion amidst the germinal layers. It is continuous anteriorly with the neural epiblast and the chorda, laterally and posteriorly with the mesoblast of the surrounding areas. It becomes perforated by an evanescent neurenteric canal and by a permanent anal orifice.

In the Cyclostomata the posterior part of the blastopore remains open as the anus. In the ventral lip of this orifice the epiblast alone is at first differentiated, the hypoblast and the mesoblast remaining fused until a comparatively late period—Goette (15) and Kupffer (34). The blastopore is closed from before backwards according to Goette's (15) statement, and consequently there is formed in front of the anus a mass of cells called the " Teloblast " by Kupffer (34). This mass of cells remains for a time fused with both epiblast and hypoblast (Goette, pl. iv, fig. 4f), but afterwards it separates from the superficial epiblast and the hypoblast (Kupffer, pl. xxviii, fig. 28), but remains continuous in front with the chorda and neural tube. Evidently the area from the front of the anus to the posterior end of the chorda and neural tube in Petromyzon corresponds closely to the anterior portion of the primitive streak of the Sauropsida and mammals. It is never perforated by a neurenteric canal, but the absence of this passage has no important bearing, as it has probably been suppressed.

In Teleostean fishes an area of fusion is found round the lips of the blastopore (Hennegny, 18), but the fusion in front of the blastopore is much more extensive than that on the sides and posteriorly, and in its anterior part a vesicle appears—" Kupffer's vesicle." In front of Kupffer's vesicle the chorda and neural plate are fused, and the margins of the fused area are continuous with the mesoblast. Eventually the posterior portion of the blastopore is closed by the fusion of its lips (18, p. 502). The position of the anus is not definitely stated. The fused area corresponds in the Teleostei even more closely than in the Cyclostomata with the anterior portion of the primitive streak of the Sauropsida and mammals, for it is partially perforated anteriorly by Kupffer's vesicle, which is evidently situated in the position of the neurenteric canal of the higher Vertebrata.

We have before referred to Balfour's description of the condition of the posterior end of the embryro in the Elasmobranchii (p. 88), and to this we have only to add that at a later period the ventral part of the blastopore also becomes closed by fusion, and that at a subsequently later period the anus is formed as a secondary opening in the line of fusion (Schwarz 47).

The observations upon the Ganoids are not sufficiently complete to afford any definite basis for comparison, but it appears (see Balfour, vol. ii. pp. 84—86, and the extract in Hoffmann and Schwalbe's 'Jahresbericht' for 1878, p. 222) that after the segmentation and during the formation of the archenteron the blastopore becomes distinct, first at its dorsal lip and then in its whole circumference; it ultimately closes, and Salensky (44) states that the anus is produced afterwards in the situation which was first occupied by the blastopore. There is, however, no evidence as to whether the blastopore in Ganoids closes from before backwards, as in Teleostei and Elasmobranchii and Cyclostomes, or from behind forward as in the *Rana temporaria*.

The primitive streak of the Amphibia appears during the period of extension of the archenteron. It is formed by a fusion of the layers in the lips of the blastopore and by concrescence of the lips of the blastopore, either from before backwards, as in Triton (Goette, 15), or by fusion of the lateral lips of the blastopore between the neurenteric canal and the anus, arriving at its completion on the obliteration of the neurenteric canal, as in Bombinator (15) and *Amblystoma punctatum* (39) or it may be formed, as we have already shown in *Rana temporaria*, by fusion of the lips of the blastopore from behind forwards.

In Teleostomes and Cyclostomes it is formed by fusion of the blastoporic lips from before backwards, also whilst the archenteron is being formed and extended, and in Elasmobranchs by fusion of the lateral lip of the blastopore behind the neurenteric canal.

In the Sauropsida and Monotremes we have no definite proof of a primary blastoporic opening, for the aperture so named by van Beneden in the rabbit (4), and by Selenka in the opossum (50), is not yet definitely located; but the primitive streak in these Vertebrates is readily comparable with the secondary condition of the Amphibian blastopore as formed in *Rana temporaria*. It is an area of fusion of the germinal layers which afterwards becomes perforated by the formation of the anus. The appearance of the fused area at a comparatively late stage in reptiles, birds and mammals—that is, after the archenteron is well established (if we except Kupffer's views on the function of the archenteron of reptiles, according to which the primitive gut is produced by invagination)—throws but little difficulty in the way of the comparison, for it is most probably a mere heterochronous displacement associated with the precocious segregation of the hypoblast in the higher Vertebrates.

Therefore, if we use the term primitive streak in the sense in which it is used in the description of the avian blastoderm—that is, as a term which signifies an area of fusion of the blastodermic layers which is continuous laterally and posteriorly with the separated epiblast, mesoblast, and hypoblast, and is in front continuous with the neural plate and chorda, and which is perforated posteriorly by the anus, and may be perforated, more or less completely, anteriorly by the neurenteric canal,—then we are bound to admit that this region and the corresponding areas in mammalian and reptilian blastoderms are the homologues of the primitive streak of *Rana temporaria*; and conversely we are forced to conclude that as the primitive streak of *Rana temporaria* is formed by the linear fusion of the undifferentiated area in the lips of the blastopore, the primitive streak of the Sauropsida and Mammalia is homologous with the fused lip of the blastopore of Rana.

Strictly speaking, therefore, the typical primitive streak is an area which extends antero-posteriorly from the point at which the fusion of layers commences, in front of the anterior lip of the blastopore (fig. 10, *Z*), to the point at which the fusion of layers terminates behind the posterior lip of the blastopore (fig. 10, *Z'*); and laterally from the termination of the fusion on the left lip of the blastopore (fig. 10, *ZZ*), to the termination of the fusion of the layers in the right lip of the blastopore (fig. 10, *ZZ'*).

If this is the case, then it is evident that the primitive streak of Bombinator, Triton, and Petromyzon, as usually described, is not homologous with the primitive streak of *Rana temporaria*, the Sauropsida, and Mammals, but only with that portion of it which lies in front of the anus. And it is also evident that the primitive streak of Teleosteans and Elasmobranchs, after the complete closure of the blastopore, is the exact homologue of the typical primitive streak as formed in birds.

With regard to the Ganoids, it is only possible to say that they seem to closely resemble the Teleosteans. It will be noticed that we have made no reference to Amphioxus. With regard to this somewhat anomalous Vertebrate, we can only observe that so far as the observations of Hatschek (16) and Kowalevsky (32) go, there is no primitive streak formed, unless we accept in full the concrescence theory, and look upon the dorsal axial line as its representative ; but even if we adopt this theoretical conception, of which there is no positive proof, we are still impressed by the fact that the anus is neither the remains of the blastopore, nor is it formed secondarily by reopening of the fused lips of the blastopore, but it appears as a new formation below and in front of the blastoporic opening ; at least Hatschek figures it in that situation and his description is as follows :—" Die Unterbrechung der Communication zwischen Darm, und Medullarrohr erfolgt ungefähr gleichzeitig oder auch etwas später als der Durchbruch des Afters. Der After bricht ventralwärts von dieser Communications-öffnung, die den letzten Rest des Gastrulamundes repräsentirt " (16, p. 79).

The anus of the frog and many other Vertebrata bears a similar relation to the neurenteric canal, but in no other vertebrate except Amphioxus is the anus a distinctly new formation ; on the contrary, it is very evidently an aperture formed by the reopening of a temporarily closed orifice, or by perforation of the homologue of that orifice. In view of this fact, it seems very probable that future observations will modify Hatschek's results, and remove the contradiction which at present exists between Amphioxus and other Vertebrata. If, however, this proves not to be the case, it will then have to be decided whether the condition which is found in Amphioxus, or that which is so general among the other Vertebrata, is the more primitive.

The Fate of the Primitive Streak in the Frog.

In discussing the primitive streak of the frog it will be convenient to consider separately—

(I) The fate of the ventral moiety.

(II) The fate of the dorsal moiety.

Goette (15) has pointed out that in Petromyzon the homologue of the primitive streak must be the whole rim of the blastopore. Apparently, however, he does not extend the same reasoning to the case of the frog. For in that case, according to his description, and to the figure he gives, in which he indicates the position of the primitive streak by a bracket, he does not include the ventral lip of the anus.

To quote his words, "Das Homologon des Primitifstreifs ist dort die ventrale Hälfte des Schwanzes von seiner Spitze bis zum After" (Haut, ventrale Hälfte des Schwanzdarms und der Mesoderm platten, Hinterwand des Aftersdarms).

It is a question how far we ought to speak of the derivatives of the primitive streak being " das Homologon " of the primitive streak. Is it quite correct to say that the adult frog is the homologue of the egg from which it has been developed ?

However, it is clear that Goette means that the ventral half of the tail is derived from the primitive streak. Here we must again differ from him.

As the result of our observations *we conclude that not only the ventral half of the tail is derived from the primitive streak, but also the dorsal half as well, with the exception of nearly the whole of the skin of the tail.*

This we shall hope to show while considering the fate of the dorsal moiety of the primitive streak.

We now proceed to describe the fate of the ventral moiety subsequent to the stage corresponding to figs. 23, 15, 16, 17.

Figs. 20, 21, 22, are camera drawings of sections of the posterior end of a 4½ mm. tadpole, taken at levels corresponding to those of sections 15, 16, and 17 respectively (*vide* lines 20, 21, 22 in fig. 26, and lines 15, 16, 17 in fig. 23).

In each case the drawings have been made by camera, and from unstained sections, and the natural appearance as regards tint and shape has been reproduced as nearly as we were able to reproduce it.

In figs. 20, 21, and 22, there is no trace of a primitive streak, but all three germinal layers are now distinct and separate.   In fig. 22 the mesoblast *ME* may be said to be   fused with the epiblast *EP'''*, but it is nevertheless, quite distinct in character of cell, and in degree of   .   In the two other sections, figs. 21 and 20, the mesoblast has   entirely separated from the epiblast.

By comparing these three sections with the sections from which figs. 15, 16, 17 were drawn, the actual fate of the primitive streak in this region may be fairly easily determined.

In comparing fig. 20 with fig. 15, the change that has taken place in the former seems to be of this nature.   The three layers, epiblast, mesoblast, hypoblast, instead of being "fused," are now (fig. 20) easily distinguishable and completely apart.

This fusion of layers we must most probably interpret as indicating an area of proliferation of cells which indeed is a characteristic feature of a primitive streak ; and when this proliferation of cells ceases, the primitive streak may be said to be no longer of physiological importance, though the area of its former extension, being of morphological interest, should be noted.

Thus, if we are regarding the primitive streaks from a physiological point of view, we must say that this portion has now ceased to exist. If we regard it from a morphological point of view, we may say that this portion has ceased to be "functional," but, nevertheless, includes the area in the bracket labelled *PS* in fig. 26 as being within the limits of the original extension of the primitive streak.   This portion we have attempted to render more distinct in fig. 26 by a different method of shading and crossed lines, *PS' PS''*.

The fate of the ventral moiety may be understood by reference to fig. 26, combined with a comparison of figs. 20, 21, 22 with figs. 15, 16, 17.   By words we may explain its fate by saying that the ventral moiety of the primitive streak splits up into portions of the three germinal layers.

(I)   Laterally and ventrally into (*a*) the posterior extremity of the mesodermal plate.

(II)   In the median plane into the two layers, (*b*) an outer or epiblastic layer, forming part of the skin, and (*c*) an inner or hypoblastic layer of large darkly pigmented cells, fig. 20, *HY'*, forming the hind wall of the rectal spout.

The difference in character of cell between the posterior wall and anterior wall of the rectal spout is very marked, as shown especially well in fig. 20. This feature is not so marked in stained specimens, where the degree of pigmentation does not show up so well.

This description so far agrees with Goette (15), and Schwarz (47), who follows Goette in this respect, except that we include the ventral lip of the anus within the primitive streak, which apparently these authors do not, though for what reasons it is not easy to understand.

Fate of the Dorsal Moiety of the Primitive Streak, and Development of the Tail.

In the dorsal moiety the fate of the primitive streak is different.

Instead of splitting up, it remains as a proliferating area; or according to our definition above, this portion of the primitive streak remains for a longer time "functional."

The relation of the neural folds to the blastopore, that is to say to the primitive streak, must be necessarily considered in connection with the fate of the dorsal moiety of the primitive streak.

Reference to fig. 14, which passes through the rapidly closing blastopore (*BL'*), conclusively proves that there is here no trace of *neural* folds. The epiblast shows no signs whatever of any thickening, the mass of cells being merely the fused layers—that is to say, the undifferentiated cells at the lips of the blastopore—of typical primitive streak.

There is no trace of neural folds—that is, of thickened epiblast separate from mesoblast—until several sections anterior to the still open portion of the blastopore. That is to say, that which from a surface examination appears to be the lower end of the neural folds, is really the dorsal portion of the lateral lips of the blastopore folded up along with the true neural folds.

If this were not so, there would be a sudden break at the posterior end of the neural folds; and the neural canal, if it opened anywhere, would open to the exterior dorsal to the blastopore, and not into the blastoporic canal, as is well known to be the actual case.

This we have tried to make clear by the diagrams, fig. 18 and fig. 19. In these diagrams the external surface of the neural plate has been dotted *NP*. The space *BR* between the two small concentric circles represents the position of the fused layers—that is to say, the lips of the blastopore at the time of the first formation of the blastopore.

The shaded space *BR'* represents the position of the same fused layers in fig. 18 at the time of the complete formation of the neural plate, but before it has commenced to fold up ; in fig. 19 after the neural plate has become completely folded, and its outer lateral edges are just meeting.

BL' is that portion of the blastopore which remains open longest.

The two asterisks in fig. 18 mark the two lines of epiblast immediately adjoining the lateral edges of the neural plate.

When the neural plate has become folded, and when its lateral edges have met and fused, and separated from the adjoining epiblast, the lines of epiblast indicated by the asterisk will have fused and made good the gap which would otherwise be caused in the skin by the separation from it of the neural plate.

In fig. 19 the folds are represented as having nearly met, so that the outer surface (dotted) of the neural plate is now no longer seen, except a narrow strip through the now rapidly approaching lateral edges of the neural plate.

The daggers similarly mark two spots in the epiblast adjoining the fused layers at the outer edges of the lateral lips of the still open portion of the blastopore.

In fig. 18 the relation of the neural plate to the dorsal lip of the blastopore—that is to say, to the dorsal end of the primitive streak—is clearly represented

In fig. 19 the folding up of the neural plate is nearly complete, and with it the dorsal part of the primitive streak—that is to say, the dorsal portions of the lateral lips of the blastopore have been folded up along with the neural folds, and the adjoining areas of epiblast marked by the daggers will shortly meet and fuse, just as will the areas of epiblast adjoining the lateral edges of the neural plate marked by the asterisks.

Similarly, just after the meeting and fusing of the epiblast along the lateral edges of the neural plate has taken place, and just as the neural plate—or tube, as it now will be—separates from and lies within the skin, so also will that portion of the primitive streak separate from the skin and come to lie within the embryo.

H

The posterior or ventral portion of the primitive streak does not become folded in this way, for by the time the folding of the dorsal portion has been completed and the separation from the skin has taken place, the lower or ventral portion has, as we have described above, split up and ceased to exist as a " functional " primitive streak.

By this means the primitive streak loses all direct connection with the surface epiblast, though the connection of the primitive streak with the surface epiblast may be said to be in reality retained by the direct continuity of the cells of the primitive streak with the interior lining (i.e. epidermic epiblast) of the neural tube.

The semi-diagrammatic sagittal sections, figs. 24, 25, and 26, will aid in rendering clearer the above statements.

Fig. 24 is an early stage, only a little older than the stage represented by fig. 18.

The " primitive streak," or area of fused layer which is cut through in these three sections, is shown by a diagonal shading.

In fig. 24 the mass of primitive streak is seen to be much greater ventrally than dorsally ; the extent of closure of the original blastopore is expressed by the distance between x and the inner edge of the lip bounding ventrally the still open portion of the blastopore.

In fig. 25 the neural plate has become folded upon itself and fused, but the neural tube so formed has not yet separated from the skin,— it is, in fact, on the point of so doing ; it therefore represents a stage rather later than that of fig. 19.

Ventrally, the primitive streak is in the middle line divided into a post-anal (PS) and pre-anal portion (PS″) by the deepening of the primitive groove at one spot (A), which shortly after completely perforates and forms the permanent anus. As we have stated above, we regard this as practically a reopening of the ventral part of the original blastopore. This is also shown diagrammatically in fig. 19, A.

Dorsally, we get a portion of primitive streak (PS″) continuous with the dorsal wall of the archenteron (HY), the notochord (CH), and floor of the neural tube, which is the actual dorsal lip of the blastopore, and is exactly the same as that portion of primitive streak which bears the same relations to the same structures—dorsal wall of archenteron (HY), notochord (CH), floor of neural tube (NP)—in fig. 24.

Externally to this and posteriorly the section cuts another portion of primitive streak (*PS"*), which is the dorsal part of the lateral lips of blastopore which have been folded up along with the neural folds, as we have already described, and as we have endeavoured to represent diagrammatically in fig. 19.

By this means a canal is formed leading from the archenteron to the neural tube, known as the *neurenteric canal, which is therefore the anterior portion of the blastopore (B'), together with the canal (PSC) bounded ventrally by the anterior portion of the primitive streak (PS'), and laterally and dorsally by the folded-over anterior portions of the lateral lips of the blastopore (PS").*

At this period there is still a direct passage to the exterior from the archenteron, as well as into the neural tube ; but as the folding up of the dorsal portion of the primitive streak progresses, combined probably with the continued concrescence of the more ventral portion of the lateral lips of the blastopore, this external opening—that is to say, that portion of the original blastopore which apparently remains open for the longest period—is entirely and finally closed from the exterior.

By the time this has occurred, or shortly after, the neural tube has become separated from the skin, and with it the dorsal portion of the primitive streak.

The primitive streak should, however, still be connected with the skin ventrally by that portion which does not become folded up and nipped off as does the dorsal portion, together with the neural tube.

Possibly this connection may exist for a very short time, but practically the separation from the skin of the dorsal moiety takes place contemporaneously with the splitting up of the ventral moiety, so that the space between the dorsal moiety of the primitive streak and the skin from which it has separated, and the space between the posterior wall of the rectal spout and post-anal gut on the one hand, and the skin at the same level caused by the splitting up of the primitive streak of that area, becomes confluent at the point in fig. 26 just where the line *PS'* crosses the space between primitive streak and skin.

Thus it comes about that the dorsal portion of the primitive streak, which remains " functional " and gives rise to the greater part of the tail, comes to lie entirely within the embryo.

Fig. 26 represents the stage at which the tail has definitely begun to grow. *The dorsal moiety of the primitive streak is seen to be lying within the embryo, and it is by proliferation of its cells that the whole of the tail is formed with the exception of the skin.* In no section after (the stage shortly antecedent to this) have we been able to see a *fusion* between the skin and the primitive streak, though the skin lies closely over the primitive streak.

The skin would seem to grow, not at any one point, but over its whole surface, on account of the pressure caused from within by the growth of the main axis of the body, in response to which the skin must either grow or rupture.

To a certain extent the skin of the middle line of the ventral surface of the tail is derived from the primitive streak; that is from the ventral moiety by the splitting up of the latter. This portion is distinguished in fig. 26 by a different mode of shading; but with the exception of this, which we think extends only a short way, no part of the skin of the tail is derived from the primitive streak.

The neurenteric canal is undoubtedly to be regarded as the most dorsal part of the blastopore; and although it remains open for a short time after the commencement of the tail, it gradually closes, as we have shown in the semi-diagrammatic figure, fig. 26, NU. This closing may, perhaps, be not incorrectly spoken of as the continuation and completion of the concrescence from behind forwards which caused the closure of the ventral portion of the original blastopore.

As far as we have been able to observe, the post-anal gut only exists during the persistence of the neurenteric canal.

In fig. 26 the part of primitive streak PS'' has been drawn too large proportionately to the part PS'.

Conclusions as regards the Primitive Streak; its Origin and Fate.

A structure exactly comparable to the primitive streak in the chick, median and grooved, is formed in the frog (*Rana temporaria*), by concrescence of the lips of the blastopore from behind forwards.

The anus perforates the posterior or ventral end of the primitive streak, being a deepening of the primitive groove. It may, therefore, be regarded as the reopening of the most ventral part of the blastopore.

The portion of the primitive streak with which the dorsal wall of the archenteron, notochord, and floor of the neural tube are continuous in the most anterior limit of the primitive streak, and is the dorsal or most anterior lip of the blastopore.

The neurenteric canal, which is bounded anteriorly by the dorsal lip of the blastopore, is therefore the most anterior portion of the blastopore.

The ventral moiety of the primitive streak shortly after the perforation of the anus ceases to exist, or, as we have preferred to term it, ceases to be "functional," and splits up.

The dorsal moiety of the primitive streak becomes folded upon itself like, and along with, the neural plate, and becomes separated from the skin, and remaining "functional," gives rise to the whole of the tail with the exception of the greater part of the skin.

We cannot find at any time any trace of blastopore or primitive streak anterior to *any part of the neural plate or tube.*

The Formation and Separation of the Mesoblast.

After the transformation of the superficial layer of the yolk-cells into epiblast is completed the remainder of the yolk may be looked upon as hypoblast and modified hypoblast, for it eventually becomes transformed, partly into the true hypoblastic lining of the enteric cavity, and partly into mesoblast and chorda. It is separated from the epiblast, except at the margins of the blastopore, and it never again, in *Rana temporaria,* becomes fused with the epiblast in front of the blastopore, after the manner described by Schultze (15) in *Rana fusca.*

During the completion of the archenteron, and before the blastopore is closed, the mesoblast, in front of the blastopore, begins to separate from the true hypoblast along the dorsolateral aspects of the archenteron by a process of delamination.

A slit-like space, which rapidly distends, appears between the hypoblast and mesoblast in these regions (fig. 12, *H*), whence it gradually extends both towards the dorsal and towards the ventral middle lines; but considerably before these clefts reach near the mid-dorsal line two sagittal clefts appear. The latter clefts separate the chorda in the middle line from the mesoblastic plates laterally (fig. 12, *S*, and fig. 6, *S*).

At this period the chorda is continuous with the hypoblast, and the mesoblastic plates and the hypoblast are still fused just outside the chorda, as well as on the ventral aspect.

Along the lines of attachment of the mesoblastic plates to the hypoblast, at the sides of the chorda, slight depressions are noticeable (fig. 6, *D*), which appear to indicate a continuation of the archenteric cavity into the mesoblast in the manner suggested by O. Hertwig (21)

Eventually the mesoblast in front of the blastopore is entirely separated from the hypoblast, first dorsally and then ventrally, the ventral fusion remaining in the region of the liver for a considerable time.

These changes do not take place simultaneously, but appear to proceed from before backwards, so that the mesoblast is still adherent to the chorda and the hypoblast, in the posterior portion of the embryo, for a short time after the separation has been completed more anteriorly.

The Primitive Streak Mesoblast.

The fusion of all the germinal layers in the lips of the blastopore continues after the margins of the orifice have concresced, and there is, therefore, behind the neurenteric canal an axial rod of tissue, which is grooved upon its upper surface, and which is not distinguishable into definite layers (fig. 15).

We have not been able to find in *Rana temporaria*, before the formation of the anus, any separation of the constituent parts of this axial rod into layers in the manner described and figured by Erlanger (10a). On the contrary, we find that the anus is formed as a perforation through the fused mass. But, after the separation of mesoblast from hypoblast has extended as far as the margins of the primitive streak, we find that the lower cells of the mesoblast are more loosely arranged, and that they are of more rounded form than the more superficial cells. When these lower more rounded cells are followed towards the streak they are seen to be continuous with the hypoblast of the primitive streak. The upper and denser layer of the mesoblast, traced in the same direction, is found to be continuous with epiblast (fig. 15).

These appearances suggest the idea that the mesoblast of this region is formed partly from the epiblast and partly from the hypoblast, but there is no definite proof that such a double formation actually occurs.

The Chorda Dorsalis.

We have already stated that in *Rana temporaria* the chorda is formed from the yolk-cells which lie beneath the neural groove. It is unimportant whether we consider these cells to be mesoblastic or hypoblastic, but we wish to emphasise the fact that the chorda in *Rana temporaria* is not formed from a layer of mesoblast which has previously separated from the hypoblast. We have been unable to find in *Rana temporaria* any appearances which would justify the conclusions which Schultze forms from his observations upon *Rana fusca*, i.e., "Dass schon auf dem Stadium der beginnenden Gastrulation in der dorsalen Urdarmwand drei Keimblätter existiren" (15).

At a later stage, however, just about the time when the medullary groove is only appearing in the posterior part of the embryo, we have noted, under the low power, an apparent separation of the mesoblast as a distinct layer across the mid-dorsal line; but this appearance we have only seen in ova which have been treated with staining reagents, and on examination with higher powers we have never been able to convince ourselves that the line of separation was normal, for we have invariably found in the space between the mesoblast and hypoblast broken fragments of cell, and we are therefore inclined to the belief that the separation of the layers was in these cases artificial, and that it had been produced by the reagents used in staining.

In unstained specimens the appearances seen in the mid-dorsal line at the time of the appearance of the medullary furrow are represented in fig. 12. The epiblast and mesoblast are separated by a distinct space (*E*). Laterally the hypoblast and mesoblast are also distinct from each other (*H*), but along the dorsal wall of the archenteron the mesoblast is neither separated from the chorda, though there are traces of the commencement of the separation on the right side (*S*), nor from the hypoblast, and the chorda is fused both with mesoblast and hypoblast.

At a later period the mesoblastic plates become entirely separated from the dorsal hypoblast, and afterwards the chorda is also separated; it then forms a distinct rod, which lies free below the floor of the neural groove and above the dorsal hypoblast except at its posterior extremity, where it terminates in the mass of cells which form the anterior wall of the neurenteric canal.

Before the commencement of the separation of the chorda from the hypoblast there are, here and there, appearances which indicate a projection of the archenteric cavity into the mass of chorda-cells, in some cases merely as a cleft-like space, in others as a more distinct diverticulum.

In *Rana fusca* O. Schultze found a peculiar secondary fusion of the mesoblast with the epiblast in front of the dorsal lip of the blastopore, "bildet sich gegen Ende der Gastrulation von der Mitte der dorsalen Urmundlippe aus eine nach dem Kopfe hin vorwärts schreitende lineare Verwachsung des äusseren und mittleren Blattes aus" (15). This line of fusion he compares to the primitive streak.

We have already said that in *Rana temporaria* there is no true layer of mesoblast along the dorsal axial line in front of the blastopore, and that in this situation the chorda is formed by differentiation of the yolk-cells ; it follows, therefore, that if the fusion occurs in Rana, it will be between the last-mentioned cells and the epiblast.

We have examined our sections carefully with reference to this point, and we find that the epiblast after its separation from the yolk-cells does not again fuse with the subjacent layer in the dorsal axial line in front of the blastopore.

It will be remembered that the separation of the blastopore terminates at a distance of ·352 mm. (p. 462) in front of the dorsal lip of the blastopore, that is, at the commencement of the primitive streak. In other words, the dorsal lip of the blastopore has an antero-posterior length of ·352 mm., and during the early stages this length does not increase ; therefore there is no forward extension of the primitive streak in front of the blastopore.

The mere fact that in *Rana temporaria* the distance from the dorsal margin of the blastopore to the point at which the epiblast is separated as a distinct layer remains constant throughout the earlier stages is not a proof that the blastoporic lips undergo no concrescence from before backwards, but it is a proof that the epiblast after its separation does not again fuse with the subjacent layer.

The Formation of the Cœlom.

Along the line of attachment of the mesoblastic plates to the hypoblast, at the sides of the chorda, it is possible to find at irregular intervals small evaginations from the archenteron (fig. 6, *D*), but these small diverticula cannot be traced for any great distance into the mesoblastic plates.

They remain for a time as blind diverticula, and then they disappear. So far as we have been able to discover they do not communicate with the cœlom, and the cœlom is not formed by extension of the archenteric diverticula, as in Amphioxus, but by a splitting of the mesoblast which first occurs laterally, and then extends dorsally and ventrally as in the higher Vertebrata.

ALPHABETICAL LIST OF LITERATURE REFERRED TO.

1. BALFOUR, F. M.—'Comparative Embryology,' vol. ii, London, 1881.

2. BALFOUR, F. M., AND DEIGHTON, F.—"A Renewed Study of the Germinal Layers of the Chick," 'Quart. Journ. Micr. Sci.,' N. S., vol. xxii, pp. 176—188.

3 VAN BENEDEN, E.—"Untersuchungen über die Blätterbildung des Chordakanals und die Gastrulation bei den Saugetieren, Kaninchen und Vespertilio murinus betreffend," 'Anatomischer Anzeiger,' Jahr. iii, pp. 709—714.

4. "Développement embryonaire des Mammifères," 'Bulletin de l'Académie Belgique,' 1875.

5. BONNET, R.—"Beiträge zur Embryologie der Weiderkäuer, gewonnen am Schafei," 'Archiv für Anatomie und Physiologie—Anatomische Abtheilung,' 1884, pp. 170—230, plates ix—xi.

6. Ibid, 1889, pp. 1—106, plates i—vi.

7. "Ueber die Entwickelung der Allantois und die Bildung des Afters bei den Wiederkäuern und über die Bedeutung der Primitivrinne und des Primitivstreifs bei den Embryonen der Saugetiere," 'Anatomischer Anzeiger,' Jahrg. iii, 1888, pp. 105—126.

8. CALBERLA, E.—"Zur Entwickelung der Medullarrohres und der Chorda dorsalis der Teleostier und der Petromyzonten," pp 226—270, plates xii and xii, 'Morphologisches Jahrbuch,' 1877.

9. DURHAM.—Notes on the Presence of a Neurenteric Canal in Rana," 'Quart. Journ. Micr. Sci.,' N. S., 1886, pp. 509, 510, plate xxvii.

10. DUVAL, M.—"De la formation du blastoderm dans l'œuf d'oiseau," 'Annales des Sciences Naturelles—Zoologie,' vol. xviii, 1884, pp. 1—208.

10A. ERLANGER, R.—"Ueber den Blastoporus der anuren Amphibien, sein Schicksal und seine Beziehungen zum bleibenden After," 'Zoologische Jahrbucher,' 1890, pp. 239—256, plates xv and xvi.

11. FLEISCHMANN, A.—"Embryologische Untersuchungen—Erstes Heft," 'Einheimische Raubtiere' Wiesbaden, 1889.

12. GASSER.—" Beiträge zur Kenntniss der Vögel—Keimscheibe," 'Archiv
    für Anatomie und Physiologie—Anatomische Abtheilung,' 1892, pp.
    359—398.

13. " Die Entstehung der Cloakenöffnung bei Hühnerembryonen," ibid., 1880,
    pp. 297—319, plates xii and xiii.

14. GOETTE, A.—' Entwickelungsgeschichte der Unke,' Leipzig, 1875.

15. ' Abhandlungen zur Entwickelungsgeschichte der Tiere, fünftes Heft
    Entwickelungsgeschichte des Flussneunauges (Petromyzon fluviatilis),'
    Leipzig, 1890.

16. HATSCHEK, B.—' Studien über Entwickelung des Amphioxus,' Wien, 1881.

17. HEAPE, W.—" Development of the Mole (Talpa Europea)," ' Quart. Journ.
    Micr. Sci.,' N. S., 1883, pp. 112—152, plates xxviii—xxxi; ibid., 1886,
    pp. 123—156, plates xi.—xiii.

18. HENNEGUY, F.—" Recherches sur le développement des Poissons Osseux,"
    ' Journal de l'Anatomie et de la Physiologie,' 1888, vol. xxiv, pp. 413—502,
    and 525—618, plates xviii—xxi.

20. HENSEN, V.—" Beobachtungen über die Befruchtung und Entwickelung
    des Kaninchens und Meerschweinchens," ' Zeitschrift für Anatomie und
    Entwickelungsgeschichte,' vol. i, 1876, pp. 213—273, plates viii and ix,
    and pp. 353—423, plates x—xii.

21. HERTWIG, O.—' Lehrbuch der Entwickelungsgeschichte des Menschen und
    der Wirbelthiere,' Jena, 1888.

22. HIS, W.—" Untersuchungen über die Entwickelung von Knochenfischen
    besonders über diejenige des Salmens," ' Zeitschrift für Anatomie und
    Entwickelungsgeschichte,' 1876, pp. 1—40, plates i and ii.

23. " Ueber die Bildung der Haifischembryonen," ' Zeitschrift für Anatomie
    und Entwickelungsgeschichte,' 1877, pp. 108—124, plate viii.

24. HOFFMANN, C. K.—" WeitereUntersuchungen zur Entwickelungsgeschichte
    der Reptilien," ' Morphologisches Jahrbuch,' 1886, pp. 176—218.

25. HUBRECHT, A. A. W.—" Studies in Mammalian Embryology. The Develop-
    ment of the Germinal Layers of Sorex vulgaris," Quart. Journ. Micr.
    Sci.,' N. S., 1890, pp. 499—562, plates xxxvi—xlii.

26. HOUSSAY, F.—" Études d'Embryologie sur les Vertébrés," 'Archives de
    Zoologie Expérimentale,' 1890, No. 2, pp. 145—215, plates x—xiv.

27. JOHNSON, A.—" On the Fate of the Blastopore and the Presence of a
    Primitive Streak in the Newt," ' Quart. Journ. Micr. Sci.,' N. S. 1884,
    pp. 659—672.

28. JOHNSON, A., AND SHELDON, L.—" Notes on the Development of the
    Newt," ' Quart. Journ. Micr. Sci.,' N. S., 1886, pp. 573—590, plates
    xxxiv—xxxvi.

29. KEIBEL, F.—" Die Entwickelungsvorgänge am hinteren Ende des
    Meerschweinchenembryos," 'Archiv für Anatomie und Physiologie—
    Anatomische Abtheilung,' 1888, pp. 407—430, plates xxiii and xxiv.

30. KOLLER, C.—"Untersuchungen uber die Blätterbildung im Huhner-keim," 'Archiv für Mikroskopische Anatomie,' 1882, pp. 174—211, plates x – xii.

31. KÖLLIKER.—'Entwickelungsgeschichte des Menchens und der höheren Thiere,' Leipzig, 1879.

32. KOWALEVSKY.—"Weitere Studien über die Entwickelungsgeschichte des Amphioxus lanceolatus," 'Archiv für Mikroskopische Anatomie,' 1877, xiii, pp. 181—204, plates xv and xvi.

33. KUPFFER, C.—Die Gastrulation an den meroblastischen Eiern," 'Archiv für Anatomie und Physiologie—Anatomische Abtheilung,' 1882, pp. 1—30, plates i—iv.

34. " Die Entwickelung von Petromyzon Planeri," 'Archiv für Mikroskopische Anatomie,' 1890, pp. 469—558, plates xxvii—xxxii.

35. LIEBERKUHN.—"Über der Chorda bei Säugethieren," 'Archiv für Anatomie und Physiologie—Anatomische Abtheilung,' 1882, pp. 401—438, plates xx and xxi.

36. MINOT, C. S.—"The Concrescence Theory of the Vertebrate Embryo," 'The American Naturalist,' 1890, pp 501—516, 617—623, 702—718.

37. MITSUKURI, K., and ISHIKAWA, C.—"On the Formation of the Germinal Layers in Chelonia," 'Quart. Journ Micr. Sci.,' N. S., vol. xxvii, 1886, pp. 17—48.

38. MOQUIN TANDON, G.—"Recherches sur les premières phases du déve-loppement des Batraciens Anoures," 'Annales des Sciences Naturelles—Zoologie,' Série 6me, tome iii, 1876.

39. MORGAN, T. H.—"On the Amphibian Blastopore," 'Studies from the Biological Laboratory of the Johns Hopkins University, Baltimore,' vol. iv., No. 6, pp. 355—377, plates xl—xlii.

40. PERENYI, J.—"Die Entwickelung der Keimblätter und der Chorda in neuer Beleuchtung," 'Anatomischer Anzeiger, Jahrg.,' iv, pp. 587—592

41. RABL, C.—"Theorie des Mesoderms," 'Morphologisches Jahrbuch,' 1889, pp. 113—252, plates vii—x.

42. RAUBER, A.—"Primitivrinne und Urmund," 'Morphologisches Jahrbuch,' Band ii, 1876, pp. 550—576, plates xxxvii and xxxviii.

43. ROBINSON, A.—'Observations on the Development of Rodents': a Thesis presented to the University of Edinburgh, 1890.

44. SALENSKY.—" Entwickelungsgeschichte der Sterlets," 1877 and 1878 ; see 'Jahresberichte der Anatomie und Physiologie,' von Hoffmann, F., und Schwalbe, G., 1878, pp. 218—225.

45. SCHULTZE, O.—"Die Entwickelung der Keimblätter und der Chorda dorsalis von Rana fusca," 'Zeitschrift für wissenschaftliche Zoologie,' xlvii, 1888, pp. 325—352, plates xxxviii and xxxix.

46. " Beitrag zur Entwickelungsgeschichte der Batrachier," 'Archiv für mikroskopische Anatomie,' vol. xxiii, 1883, pp. 1—22, plate i.

47. Schwarz, D.—" Untersuchungen der Schwanzendes bei den Embryonen der Wirbelthiere," 'Zeitschrift für wissenschaftliche Zoologie,' xlviii, 1889, pp. 191—223, plates xii—xiv.

48. Scott, W. B., and Osborn, H. F.—"On some Points in the Early Development of the Common Newt," 'Quart. Journ. Micr. Sci.,' N. S., 1879, pp. 449—472, plates xx and xxi.

49. Sedgwick, A.—"On the Origin of Metameric Segmentation, and some other Morphological Questions," 'Quart. Journ. Micr. Sci.,' N. S., 1884, pp. 24—42, plates ii and iii.

50. Selenka, E.—'Studien über Entwickelungsgeschichte der Thiere.—Das Opossum,' Wiesbaden, 1886.

51. Sidebotham, H.—"Notes on the Fate of the Blastopore in *Rana temporaria*," 'Quart. Journ. Micr. Sci.,' N. S., 1888, pp. 49—54, plate v

52. Spencer, W. B.—" Some Notes on the Early Development of *Rana temporaria*,"; 'Quart. Journ. Micr. Sci.,' N. S., Supplement, 1885, pp. 123—137, plate x.

53. Strahl, H.—" Beiträge zur Entwickelung der Reptilien," 'Archiv für Anatomie und Physiologie—Anatomische Abtheilung.' 1883, pp. 1—13, plate i.

54. " Beiträge zur Entwickelung von *Lacerta agilis*," 'Archiv für Anatomie und Physiologie—Anatomische Abtheilung,' 1882, pp. 242—278, plates xiv and xv.

55. " Zur Bildung der Cloake des Kaninchen-embryo," 'Archiv für Anatomie und Physiologie—Anat. Abtheil.,' 1886, pp. 156—168, plate iv.

56. Weldon, R.—" Notes on the Early Development of *Lacerta muralis*." 'Quart. Journ. Micr. Sci.,' N. S., xxiii, 1883, pp. 134—150, plates iv—vi.

57. Wenckebach, K. F.—" Der Gastrulationsprocess bei *Lacerta agilis*," 'Anatomischer Anzeiger,' 1891, pp. 57—66 and 72—77.

EXPLANATION OF PLATES III & IV.

Illustrating Dr. Arthur Robinson's and Mr. Richard Assheton's paper on "The Formation and Fate of the Primitive Streak, with Observations on the Archenteron and Germinal Layers of *Rana temporaria*."

Alphabetical List of Reference Letters for all the Figures.

A. Anus. a. Anterior extremity of archenteron. A P. Post-anal gut. A R. Archenteron. A R'. That portion of the archenteron the posterior wall of which is formed by the closure of the blastopore. B, B'. Rim of blastopore lip. *b*. Dorsal lip of blastopore. *B C*. Blastoporic canal. *B L*. Blastopore. *B L'*. Portion of original blastopore not yet closed. *B R*. Original position of fused layers, that is to say, lips of blastopore at the time of the first formation of the circular blastopore. *B R'*. Position of fused layers or lips of blastopore after concrescence to a greater or less extent of the lateral lips of the original blastopore. *B R''*. The dorsal portion of the lateral lips of the blastopore folded over with the neural plate, so that which was their ventral or inner surface is now seen from without. *c*. Ventral lip of blastopore. *C H*. Notochord. *D*. Diverticulum from archenteron towards mesoblast. *d*. Line of section of Fig. 2. *E*. Space between epiblast and mesoblast. *E P*. Epiblast. *E P'*. Epidermic layer of epiblast. *E P''*. Nervous layer of epiblast. *H*. Slit between hypoblast and mesoblast. *H Y*. Hypoblast. *H Y'*. Primitive streak of hypoblast. *J*. Line of fusion of the folded-up edges of the neural plate. *M E*. Mesoblast. *M E'*. Epiblastic or outer layer of mesoblast of primitive streak. *M E''*. Hypoblastic or inner layer of mesoblast of primitive streak. *N C*. Cranial nerve. *N G*. Neural groove. *N P*. Neural plate. *N S*. Roof of spinal cord. *N T*. Central canal of spinal cord. *N U*. Neurenteric canal. *P G*. Primitive groove. *P S*. Primitive streak. *P S'*. The part of the primitive streak ventral to the anus, *i.e.* ventral lip of original blastopore. *P S''*. The part of the primitive streak which is continuous with the posterior end of the notochord, *i.e.* dorsal lip of original blastopore. *P S'''*. The part of the primitive streak which lies between that portion of the blastopore which remains open longest and the anus, *i.e.*, the greater part of the fused lateral lips of the original blastopore. *P S''''*. The part of the primitive streak which is folded up with the neural folds and forms the posterior wall of the neurenteric canal. *P S c*. The dorsal portion of the neurenteric canal. *S*. Line of separation of mesoblast from chorda. *S G*. Segmentation cavity. *x*. The lowest extremity of that part of the archenteron which is bounded posteriorly by the fused lips of the blastopore. *Y*. Yolk cells. *Y P*. Yolk-plug. *Z, Z', ZZ, ZZ'*. Furthest points from the edge of the blastopore, at which a distinct fusion of germinal layers is perceptible. * *. Two points of epiblast beyond the neural plate, which meet and fuse when the folding up of the neural plate is completed. † †. Two points of the epiblast beyond the dorsal portion of the primitive streak, which meet and fuse when that portion of the primitive streak is folded up with the neural plate.

Fig. 1.—Sagittal section of a portion of the ovum represented in Fig 4, showing the dorsal lip of the blastopore and the area of fusion in front of it. The cavity of the archenteron, and the pigmented streak in front of it which indicates the line of extension. × 50.

Fig. 2.—A transverse section through the pigmented area in front of the archenteric cavity. The section is taken in the direction of the line *d* in Fig. 1. × 115.

Fig. 3.—A sagittal section through the posterior lip of the blastopore of the ovum represented in Fig. 4. × 115.

Fig. 4.—A sagittal section of an ovum at the period of completion of the blastopore. × 23.

Fig. 5.—A sagittal section of an ovum after the disappearance of the segmentation cavity, and when the blastopore is partially closed.   × 23.

Fig. 6.—A portion of a transverse section of an an ovum after the separation of the chorda from the mesoblast, but before the complete separation of the mesoblast and chorda from the hypoblast.   × 138.

Fig. 7—A nearly horizontal section through the posterior part of an embryo in which the blastopore had only slightly diminished.   The section is taken through the centre of that part of the original blastopore which still remains open. Drawn with camera, × 30.

Fig. 8.—A nearly horizontal section through the same embryo, but taken ventrally to the existing blastopore through the ventral portion of the original blastopore, the lateral lips of which have coalesced.   × 30.

Fig. 9.—A nearly horizontal section through the same embryo, but taken ventral to both figs. 7 and 8.   It was the forty-first section ventral to that of Fig. 7.   × 30.

Fig. 10.—A diagram to illustrate the changes that had taken place in the shape of the blastopore, in the embryo of which Figs. 7, 8, and 9 are sections. The lines numbered 7, 8, 9 are drawn about the levels at which the sections bearing those numbers are taken.   The space lettered B R, enclosed by the two small concentric circles, represents the position and extent of the fused layers of the blastoporic rim at the time of the first formation of the circular blastopore or anus of Rusconi ; the shaded space B R' represents the position and extent of the fused layers at the stage of the embryo of which Figs. 7, 8, and 9 represent sections.   The line lettered x represents the extent of the closure of the ventral portion of the lateral lips up to this stage.   The letters Z, Z', ZZ, ZZ', refer to the same spots as those letters do in Figs. 5 and 7.

Fig. 11.—A surface view of the posterior end of an embryo, in which the blastopore has diminished considerably, but in which it is still circular.   The neural plate, N P, is distinctly visible.   The line PG is the primitive groove, at this period only with difficulty made out in surface view.   In the figure it is drawn too distinctly.   The levels at which the sections Figs. 12 and 13 were taken are indicated by the lines numbered 12 and 13.   These sections (Figs. 12 and 13) were not taken from the same embryo from which Fig. 11 was drawn. This drawing was made from a living embryo.

Fig. 12.—A transverse section through the posterior portion of the neural-plate region, showing the commencing separation of the mesoblast from the chorda.   × 115.

Fig. 13.—A transverse section through the posterior part of the primitive streak of an embryo, about the same age as that represented in Fig. 5.   The section is taken along the line x in the latter figure.   × 50.

Figs. 14, 15, 16, and 17 are horizontal sections through the posterior end of the same embryo, the stage being precisely the same as that of Fig. 23, on which the levels of the four sections are shown by the lines 14, 15, 16 and 17. Fig. 14 is taken through the centre of the nearly closed "blastopore," and the sections 15, 16, and 17 are the 16th, 32nd, and 44th sections respectively below section 14.

Fig. 14.   A horizontal section through the posterior end of the embryo,
and therefore a section transverse to the longitudinal axis of the
primitive streak.   The section is taken through the centre of the small
part of the original blastopore which is still open, which, however, is
seen to be on the point of closing.   The three germinal layers are
fused at the blastoporic lip, and two layers of the mesoblast may be
distinguished—a denser outer or epiblastic layer, *ME''*, and a looser
inner or hypoblastic layer, *ME'''*.   The cells of the latter are further
characterised by being rounder and more deeply pigmented.   Drawn
with camera, × 55.

Fig. 15.   2½ mm. Tadpole.   A horizontal section parallel to the preceding
section, taken through about the centre of the primitive streak.   It
presents practically the same features as the preceding figure, except
that at this point the blastopore has completely closed, and the opposite
lips have fused, forming a typical primitive streak.   Camera drawing,
× 55.

Fig. 16.   2½ mm. Tadpole.   Horizontal section through the posterior part
of the embryo.   This section passes through that portion of the
primitive streak in which the anus will shortly appear.   The primitive
groove is at this point very much deeper, forming that which is usually
called the proctodæum; but complete perforation has not yet occurred.
The line of perforation is, however, foreshadowed as it were by a deep
irregular line of pigment.   Camera drawing, × 55.

Fig. 17.   2½ mm. Tadpole.   A horizontal section of posterior end of an
embryo ventral to the anus.   This is the thirteenth section behind the
anus.   The primitive groove is still well marked.   As there is no true
hypoblast here, there is fusion only between epiblast and mesoblast.
Drawn with camera, × 55.

FIG. 18.—A diagram to show the relation of the neural plate to the
"blastopore" and rest of the primitive streak.   The dotted area represents
the neural plate.   The shaded area *B R'* represents the primitive streak at the
time just prior to the commencement of the folding up of the neural plate.
The space *B R* between the two concentric circles represents the original
position of *B R'*.   The asterisks denote two points in the epiblast which will
come together when the neural plate has folded up, and the daggers denote
two other points alongside the blastoporic lips which will come together as
explained in the text.

FIG. 19.—A diagram to show the relation of the neural plate to the blastopore
and rest of the primitive streak, and to show the way in which a portion of
primitive streak becomes folded up along with the neural folds.   It will be
seen that here the folding is nearly completed;   the asterisks and daggers of
Fig. 18 have been brought together so as to almost coincide.   This figure also
diagrammatically represents the anus as a re-opening of the ventral part of
the original blastopore.   This diagram and the diagram of Fig. 18 represent
nearly the stages of the drawings Fig. 23 and Fig. 11 respectively.

FIG. 20.—4½ mm. Tadpole.   A horizontal section through the posterior end
at a level corresponding to that of Fig. 15, but of an older embryo.   Instead

of the layers being fused along the middle line, they are now separated and distinct. Camera drawing, × 55.

FIG. 21.—4½ mm. Tadpole. A horizontal section through the posterior end at a level corresponding to that of Fig. 16, that is to say, through the anus, which is now completely formed. Camera drawing, × 55.

FIG. 22.—4¼ mm. Tadpole. A horizontal section through the posterior end at a level corresponding to that of Fig. 17. These three sections, Figs. 20, 21, and 22, are from the same series. The sections are not stained, and the drawings have been made to represent the actual shades of the three layers as nearly as possible. Camera drawing, × 55.

FIG. 23.—2½ mm. Tadpole. A surface view of the posterior end of a Tadpole, in which the neural folds have met and are fusing. The "blastopore" is still open, and the anus is conspicuous. The horizontal lines numbered 14, 15, 16, and 17 are drawn at the levels at which the sections were taken, from which Figs. 14, 15, 16, and 17 were drawn. The bracket, P S, includes the dorsal and ventral limits of the primitive streak. This figure was drawn partly from a specimen hardened in Kleinenberg's picro-sulphuric acid, and partly from a model constructed by pasting together in order pieces of card-board corresponding to a complete series of camera drawings.

FIG. 24.—A sagittal section of the posterior three-quarters of a frog embryo, in which the neural plate is beginning to fold up, and in which the anus of Rusconi has become much reduced. The yolk-plug has been retracted, and the embryo is lengthening. The extent of the primitive streak is indicated by the bracket P S, and by the diagonal shading.

FIG. 25.—2½ mm. Tadpole. A semi-diagrammatic sagittal section through the posterior half of a frog embryo, of the same age as that from which Fig. 23 was drawn. The extent of the primitive streak is indicated by the bracket P S, and by the diagonal shading. The neural plate has completely folded over, but has not yet separated from the external epiblast. This figure shows the dorsal portion of the primitive streak folded over with the neural plate, and with it about to be separated from the adjoining epiblast. The blastopore has not quite closed in the middle line. The neurenteric canal is seen to be formed of two portions; the ventral, being the original opening from archenteron to exterior, the blastoporic canal (B. C.) or blastopore, and a dorsal, the canal through the folded-up lateral lips of the blastopore, or portion of the primitive streak (P S C). The anus has not yet completely perforated the ventral part of the primitive streak.

FIG. 26.—4½ mm. Tadpole. A semi-diagrammatic sagittal section through the posterior end of an embryo in which the tail has just begun to grow out. The central nervous system has entirely separated from the skin, and with it that portion of the primitive streak which was folded up with it (P S^{br}). By this means that portion of the primitive streak (P S^{br}) comes to lie now within the embryo. The ventral moiety of the primitive streak has ceased to exist as such, and the portions derived therefrom are shown by cross-shading; while the dorsal portion, or still functional portion, if it may be so termed, is indicated as in the two immediately preceding figures by diagonal shading. The bracket P S includes the full extent of primitive streak or its derivatives.

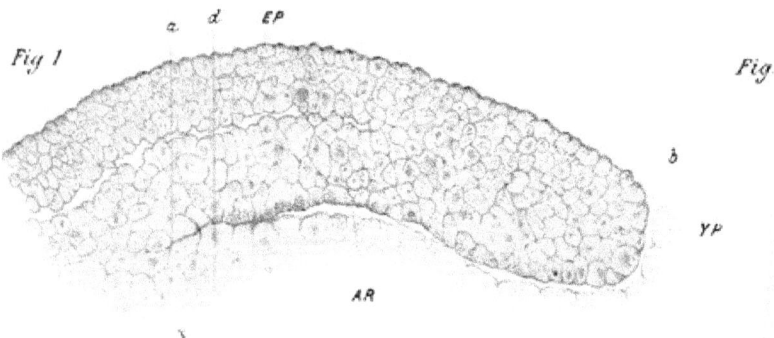

Fig 1

Fig

a. d EP

b

YP

AR

x

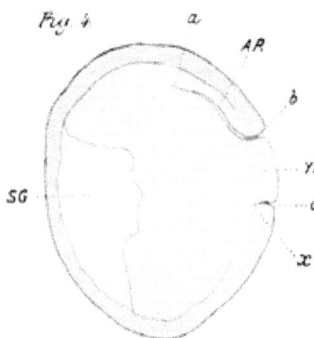

Fig. 4.

a

AR

b

YP

SG

c

x

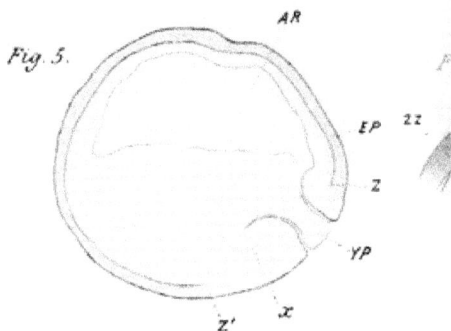

Fig. 5.

AR

F

EP

ZZ

Z

YP

Z'

x

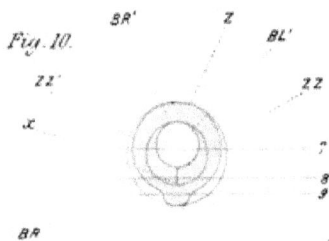

Fig. 10.

BR'

Z

BL'

ZZ'

ZZ

x

7

8

9

BR

z'

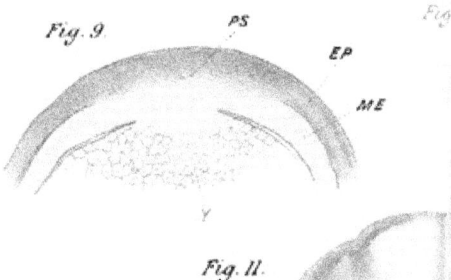

Fig. 9.

PS

EP

ME

Fig

Y

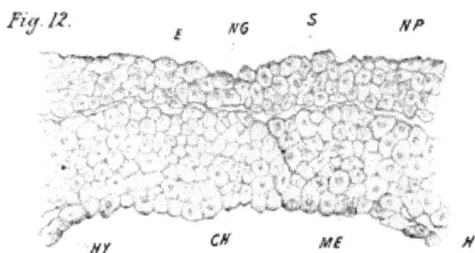

Fig. 12.

E NG S NP

NY CH ME H

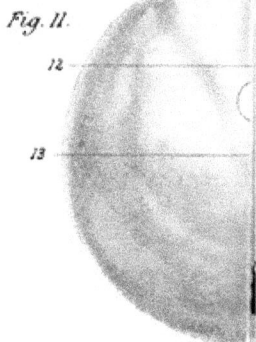

Fig. 11.

12

13

Fig. 3.

EP

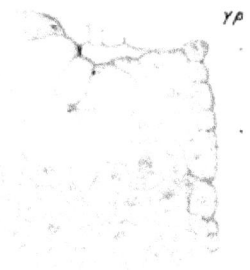

YP

Fig. 6.

EP   NP

NC   NG

AR

B   B'

ZZ

HY   EP

Y   ME

X   PS

EP   ME

ME

Y   HY   D   S

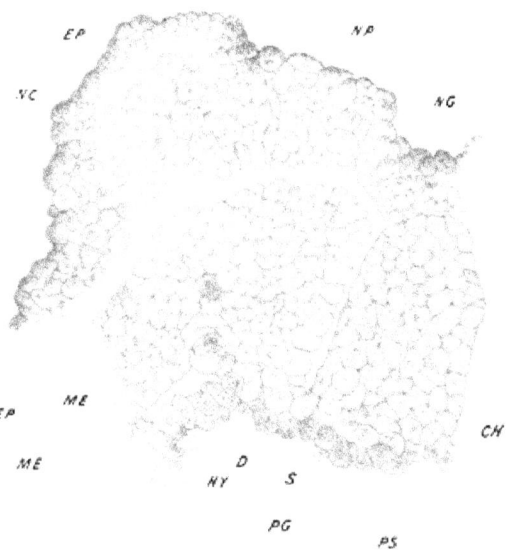

CH

PG   PS

NP

BL

Fig. 13.

EP

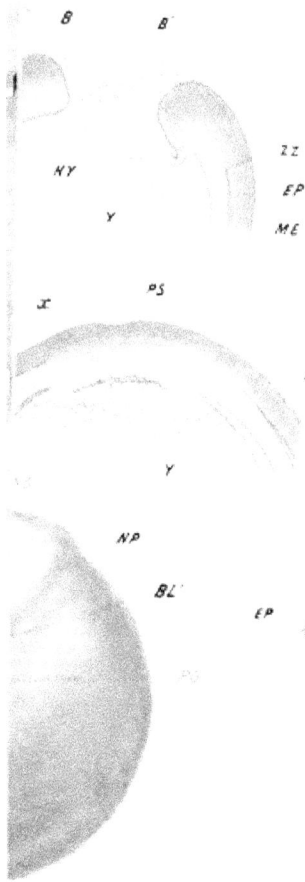

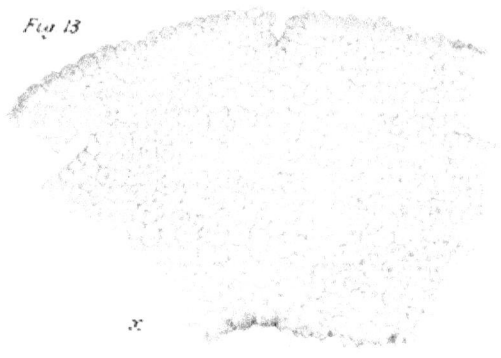

X

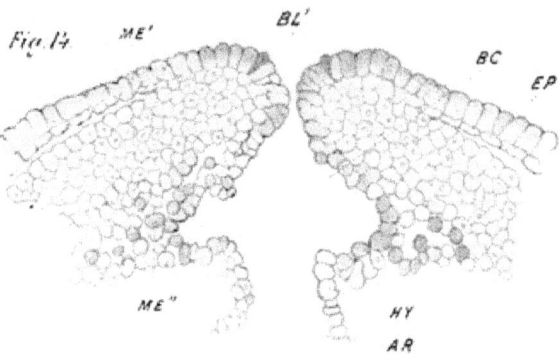

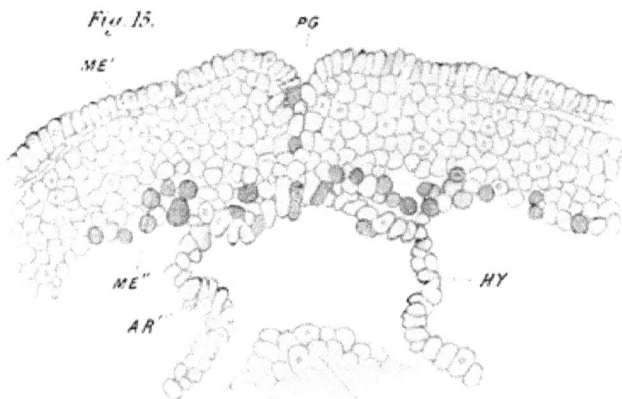

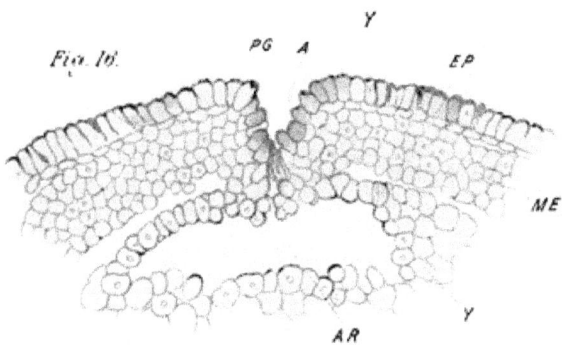

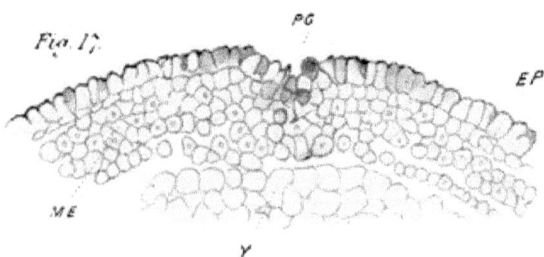

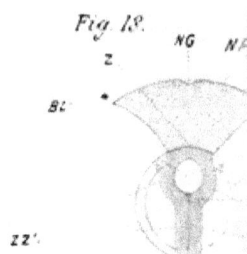

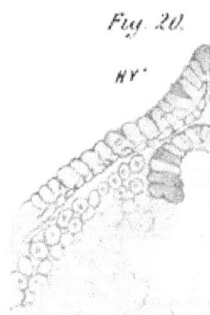

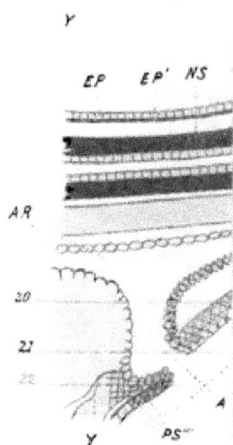

Fig. 14.
ME'
BL'
BC
EP
ME"
HY
AR

Fig. 18.
Z
NG
N.P
BL
ZZ'

Fig. 15.
PG
ME'
ME"
HY
AR'
Y

Fig. 20.
HY'
Y

Fig. 16.
PG A
EP
ME
Y
AR

EP EP' NS
AR
20
21
22
Y
PS'

Fig. 17.
PG
EP
ME
Y

Fig. 26.

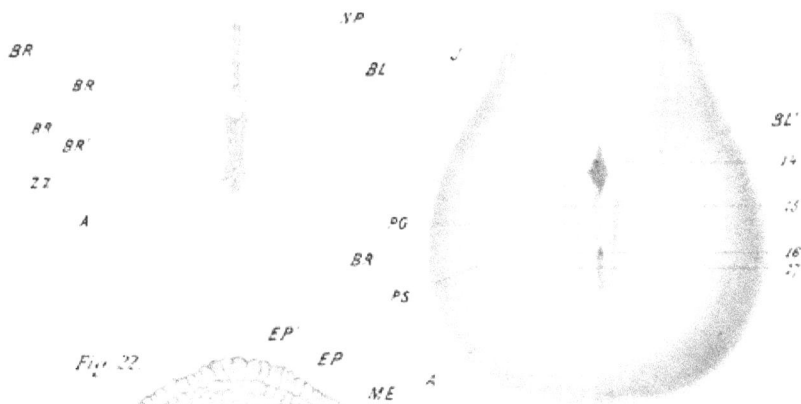

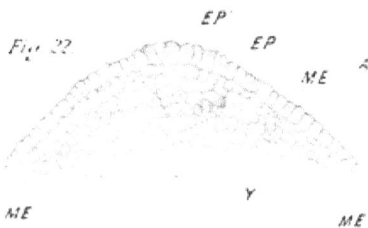

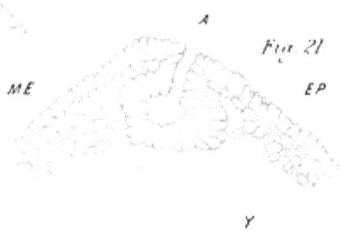

Fig. 22

Fig. 21

Fig. 23

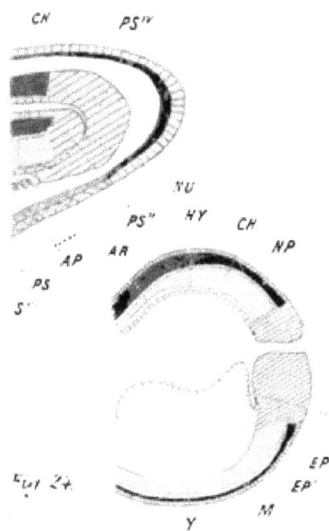

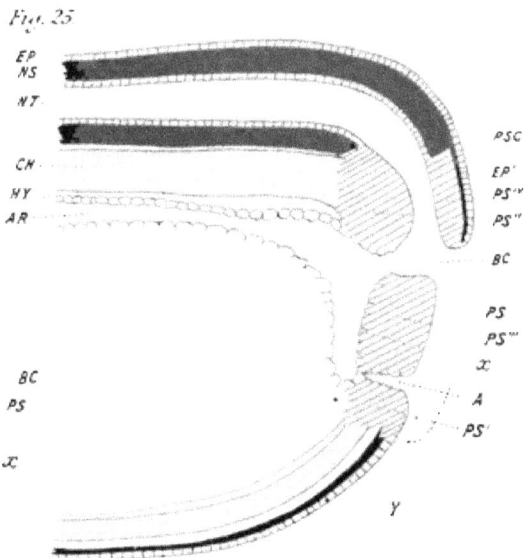

Fig. 24

Reprinted from the JOURNAL OF THE MARINE BIOLOGICAL ASSOCIATION, *New Series, Vol. II., No. 2.*

ON SOME ASCIDIANS FROM THE ISLE OF WIGHT:

A STUDY IN VARIATION AND NOMENCLATURE.

By WALTER GARSTANG, M.A., *Berkeley Fellow of the Owens College, Manchester.*

With Plates V. and VI.

Although the Isle of Wight has been a favourite haunt of the geologist and the palæontologist, references to its present marine fauna are exceedingly rare in zoological literature. Early in May of the present year, however, I had an opportunity, at the suggestion and through the kind hospitality of my friend Mr. Poulton, of examining the littoral fauna of the eastern shores of the island, and of making a considerable collection of zoological specimens. A list of the species which I obtained will be published as soon as I have had time to complete the examination of them; but several of the Ascidians throw so much light upon the brief and obscure descriptions of certain species, that I believe it will be serviceable to give a full account of them without further delay, especially since the pressure of other work may prevent an early appearance of the complete list.

I.

Ascidia mollis, *Alder and Hancock.*

ASCIDIA MOLLIS, *Hancock.* Ann. Mag. Nat. Hist. (iv), vol. vi, 1870, pp. 358, 359.

I found eleven individuals of this species attached to rocks in the *Zostera* bed off Nodes Point, St. Helen's, at extreme low water, May 7th.

The short account of *A. mollis* given by Hancock is admirable as regards the description of the external features, but is insufficient in some points of internal structure. I am glad, therefore, to have an

1

opportunity of re-describing this Ascidian. It appears to be a comparatively rare species, and I am not aware that it has been hitherto recorded from any other locality than the coast of Connemara, in the West of Ireland.

The *body* is ovate in form, thick, lobate, attached generally by the posterior half, sometimes by a larger area of the left side. When living, it is invariably of a rosy-flesh colour; and this colour, upon close examination, is seen to be due to a number of crimson dots (the culs-de-sac of the test vessels) profusely scattered in the substance of the test.

The dimensions of the largest individual (Pl. V, fig. 1) are as follows:

Maximum length (antero-posterior)   .   . $1\frac{1}{10}$ inch.
,,   breadth (dorso-ventral) .   .   . $\frac{3}{4}$,,
,,   thickness (right to left) .   .   . $\frac{3}{8}$,,

The oral and cloacal *apertures* are on the right side of the body; the oral is sub-terminal, the cloacal half-way down and near the dorsal edge; both are small and inconspicuous. The position of the cloacal aperture varies very little in these specimens; in a few it is slightly posterior to the middle dorso-ventral line, but never so much so as to be two-thirds of the way down. No *ocelli* were observed around the apertures.

The *test* is, in Hancock's words, "firm, thick, semi-transparent, smooth and soft to the touch, rather shining, obtusely lobed, of a rosy-flesh colour, showing minute punctures and veinings of crimson." In its thick, smooth, firm, and shining character, the test of this species resembles that of *Phallusia mammillata*, a resemblance further borne out by the lobes of its surface, although these are much flatter and less protuberant in *Ascidia mollis*, than in the latter species. In its softness, however, the test of this species is very unlike that of *P. mammillata*.

Hancock states that the "terminal extremities" of the blood-vessels of the test are "more inflated and globular in this than in any other species." I have a distinct recollection of their pyriform character in the living anima but their appearance in specimens after preservation in alcohol is very different; and they are seen to be elongated and finger-shaped, rather than inflated and globular (Pl. VI, fig. 2).

When the test has been removed from the rest of the body, the oral and cloacal *siphons* are seen to be short (Pl. V, fig. 3). The musculature is, as usual, almost confined to the right side of the body; the fibres are long and delicate. Round each of the siphons a number of delicate fibres form a complete sphincter.

On the left side, the course of the intestine is visible through the body-wall. The *stomach* is rounded in form and is situated at some little distance (about one-fifth of the total body-length) from the posterior end of the body. The *intestine* is narrow and uniform ; its first bend is well in front of the cloacal siphon, its anterior wall being on a level with the ganglion ; the second bend of the intestine is behind the cloacal siphon, its posterior wall being on a level with the opening of the œsophagus into the pharynx ; the rectum is directed obliquely forwards towards the cloaca.

Upon opening the pharynx from end to end, along the line of the endostyle, the remaining structures can be examined.

The *coronal tentacles* are forty or more in number. In an individual possessing forty tentacles, they were of three sizes and regularly arranged—ten long and slender primaries, ten intermediate secondaries, and twenty short tertiaries.

The *præbranchial zone* is studded with microscopic papillæ.

The aperture of the *dorsal tubercle* is crescentic in the smaller specimens, horse-shoe shaped in the individual represented in fig. 1, the horns not being incurved.

An *epipharyngeal groove* extends along one-third of the distance between the dorsal tubercle and the ganglion, which is situated half-way between the mouth and the cloacal aperture. The ganglion is small, three times as long as broad, and extends over three of the meshes of the pharyngeal wall, beginning at the fourteenth horizontal bar. The epipharyngeal groove becomes elevated towards its posterior end, and behind it commences the *dorsal lamina*, which is very narrow, strongly ribbed transversely, and pectinated at its margin. The ribs and teeth of the lamina correspond in number with the horizontal bars of the pharyngeal wall. Occasionally there are minute projections from the edge of the lamina which alternate with the teeth in position. The concave side of the lamina shows a series of weak ridges running towards its edge very obliquely from before backwards.

Branchial apparatus.—A portion of the inner face of the pharyngeal wall is shown on Pl. V. fig. 4. The horizontal vessels form three complete series and a rudimentary growth. The primary vessels (*h. v.* 1), which give off branches* to the body-walls, are usually of greater diameter than those of the remaining series. Between each pair of primaries are situated one secondary vessel (*h. v.* 2), and two tertiary vessels (*h. v.* 3), at approximately equal distances.

Connecting ducts (*c. d.*) arise from all these vessels and support delicate internal longitudinal bars (*i. l. b.*) which are surmounted at the points of junction by moderately stout conical papillæ (*p.*) and at intermediate points by comparatively slender ones (*i. p.*). The connecting ducts themselves are sub-triangular in shape when seen in profile. The horizontal and internal longitudinal vessels delimit meshes, which are sometimes almost twice as long as broad, and contain four or five stigmata each. The stigmata are elongated, with rounded ends; they are frequently double, and then consist of an anterior and a posterior portion of elliptical shape. The pharyngeal wall is minutely plicated in a longitudinal direction. The meshes almost invariably show some traces of a division into two equal portions by the formation of an incomplete quaternary series of horizontal vessels; the extent to which this process is carried out varies in different individuals and in different parts of the same pharynx. The process is interesting, and may be completely traced in fig. 4. A small projection arises from the internal face of an interstigmatic bar, at its middle point (see fig. 4, upper row, third mesh from the left), and is joined by a similar projection from the opposite wall of the stigma (see the mesh below). The concrescence of the two projections forms a horizontal bridge across the middle of the stigma. The formation of several such bridges across adjacent stigmata thus gives rise to a small horizontal vessel (see the mesh below), which may be said to form part of a quaternary series; these quaternary vessels (*h. v.* 4) may even form connections with the internal longitudinal bars beneath the intermediate papillæ (*i. p.*) of those structures. My figure represents the condition of the branchial apparatus in the individual shown in fig. 3; but in a some-

---

* The origin of these branches—the dermato-branchial connectives—is marked in some specimens by white spots upon the primary horizontal bars.

what larger individual (fig. 1) the intermediate or quaternary vessels are much more highly developed, and there is less difference between them and the other horizontal vessels.   There is no pharyngo-cloacal slit.*

The œsophagus opens into the pharynx high up on its dorsal edge, half-way between the cloacal siphon and the poterior end of the body. In the largest specimen there are six primary horizontal bars between the œsophageal opening and the posterior end of the pharynx.

All my specimens are immature ; in even the largest individual the development of the generative organs is still incomplete, and the ducts are very slender in form ; while in smaller specimens the gonads are quite rudimentary.

In addition, however, to the specimens of which the above account has been given, I took another Ascidian which there is every reason to believe to be an adult individual of the same species, but which, from its exceptional shape, is at least an abnormal one, so that I have excluded it from the general description.

It is represented of the natural size on Pl. V, figs. 5a and 5b, and deprived of the test, by fig. 6.   The body is not compressed from side to side (right to left morphologically) like an ordinary Ascidian and like normal individuals of the same species, but dorso-ventrally ; and thus it comes about that, although attached in the usual manner by its left side, its right side does not present a flattened surface, but is elevated so as to form a thickened longitudinal ridge of considerable height.

The dimensions are as follows :

Length (antero posterior)	2 inches.
Breadth (dorsal-ventral across the plane of attachment)	$\frac{5}{8}$,,
Thickness (morphological right to left)	$\frac{1}{8}$,,

The breadth becomes considerably reduced towards the summit of the ridge which represents the right side of the body.

The test is very thick and presents all the characters of normal individuals of Ascidia mollis, except that it is much corrugated on that face of the body which contains the cloacal aperture (see fig. 5a). The oral aperture is subterminal and on the same side as the cloacal

* See page 144.

aperture, which is slightly nearer the anterior than the posterior end
of the body. The body is attached by almost the whole of the left
side, which is deeply furrowed and irregularly pitted. The test is
overgrown by extensive colonies of the Polyzoon *Cylindrœcium
dilatatum*.

Upon removal of the test, the extent of the dorso-ventral compression is at once noticed. The *ganglion* and *cloacal siphon*, instead
of occupying their usual position upon the apparent left of the body,
are in the median line of the upper side; and the whole of the
viscera appear to have suffered a similar rotation through 90 degrees.
Strictly speaking, however, the viscera present exactly the same
morphological relations to the rest of the body as in the normal
individuals described above.

The *generative organs* are well developed, and the oviduct and vas
deferens are remarkably dilated. The former contains numerous ripe
ova, of small size; and the latter is filled with a mass of spermatozoa.

The only difference of any importance in the pharynx is the presence
of a *pharyngo-cloacal slit*,* $\frac{1}{8}$ inch in length, in the usual position
opposite the cloacal aperture.

The *præbranchial zone* is closely studded with minute papillæ.

The growth of the aperture of the *dorsal tubercle* has progressed
still further; both horns are now curved inwards.

Epipharyngeal groove and *dorsal lamina* as in younger specimens,
but the lamina is a little deeper; the ribs are very strong and regular,
the teeth rather short, very regular, without intermediate smaller
ones; the concave side of the lamina as described above.

Behind the œsophageal aperture is a long smooth area (the "post-
buccal raphé" of Roule), bounded on the left by a continuation of
the dorsal lamina, and on the right by a series of terminal elevations
of the horizontal membranes of that side, as in *A. mentula*.

Branchial apparatus.—This is much as in younger specimens,
but the arrangement of horizontal bars into primaries and secondaries,
&c., is less obvious, owing to the increase of the quaternary vessels
which are in many parts of the pharynx completely formed. Rudimentary quarternaries are rare.

The papillæ at the junctions are bluntly conical; the intermediate

* See p. 144.

papillæ are well developed and slenderly conical. The meshes are slightly longer than broad, except where new quaternaries are forming, when they are twice as long. There are from five to seven stigmata in a mesh.

Between this Ascidian and the immature specimens of *A. mollis* described above, the only points of difference, which are not obviously the consequences of further growth, are the different plane of compression and the presence of a pharyngo-cloacal slit.

As to the former, it is a pure abnormality. By Professor Lankester's kindness I have had an opportunity this year of examining in detail the collections of Tunicates in the Oxford Museum, and I found there a specimen of *Phallusia mammillata* which exemplified precisely the same kind of variation. The broadly ovate test was compressed dorso-ventrally, the apertures and ganglion being in the middle line of the upper side, and the viscera and visceral septum of the test being correspondingly rotated. Yet there were no structural differences at all to warrant a division of the species.

As to the pharyngo-cloacal slit, its presence in the adult and not in the young may seem surprising, especially when its supposed morphological importance is taken into account; but I have found exactly the same phenomena in the species *Ascidiella aspersa*. The ordinary specimens of that species show no trace of this aperture, but I have seen a distinct pharyngo-cloacal slit in a particularly large individual, taken from a Falmouth trawler, which I examined this year at Plymouth; in it the slit* occupied its usual position opposite the cloacal aperture. It may therefore be admitted that the presence of this slit is in some way a consequence of increased size, and that its absence in young individuals is not a matter of specific value. An attempt to explain the meaning of this remarkable aperture is made below (see p. 144).

---

* The walls of the slit were definite, straight, and smooth, resembling in all respects those of the slit in *Ascidia mentula*. It must not be imagined that the slit which I have mentioned, was an irregular abnormality of the kind described by Professor Herdman in specimens of *Ascidiella aspersa* from the west coast of Ireland (Proc. Liv. Biol. Soc., v. 1891, p. 210, pl. x), an abnormality which may also occur in *Ascidia mentula*, as I have myself observed in a specimen from Lough Long.

II.

Ascidia depressa, *Alder*.

ASCIDIA DEPRESSA, *Alder*. Cat. Moll. North. Durham, Trans. Tyneside
Nat. Field Club, 1848, p. 107.

—   —   *non Heller.* Untersuch. über die Tunicaten d.
Adriat. Meeres, Denks. d. Kais.
Akad. Wiss. Wien., xxxiv, ii, 1875,
p. 15, Taf. v., figs 10—12.

—   —   *nec Herdman.* Notes on British Tunicata, Journ.
Linn. Soc., xv., 1881, pp. 286, 287,
pl. xviii., figs. 4, 5.

—   —   *nec Roule.* Rech. s. les Ascidies Simples d. Cotes
de Provence, Ann. Mus. d'Hist.
Nat. Marseille, tom. ii, 1884.

Under this name I describe a species of Ascidian of which I took
four specimens on May 11th. They were attached to the under surface of a stone near the Zostera bed off Nodes Point.

SPECIFIC DIAGNOSIS.—*Body* oblong ovate, much depressed, greenish
when alive, attached by the whole of the left side. *Oral aperture* subterminal; *cloacal* two-thirds of the way down, on the right side, near the
dorsal edge. *Test* rather thin, cartilaginous, provided with numerous
minute tubercles on its free surface. *Oral* and *cloacal siphons*, especially
the cloacal, rather long. *Stomach* rounded, at the posterior end of the
body; first bend of intestine considerably anterior to the cloacal siphon;
rectum directed obliquely forwards, sometimes almost horizontal. *Tentacles* 25 to 30, long and slender. *Præbranchial zone* studded with minute
papillæ. Aperture of *dorsal tubercle* horse-shoe shaped, horns not incurved concavity anterior. *Ganglion* much elongated, slightly dilated at
each end. *Epipharyngeal groove* low, moderately long. *Dorsal lamina*
continued behind the œsophageal opening, fairly deep, strongly ribbed on
the convex side and regularly pectinated with stout papillæ profusely
scattered on the concave side. *Pharyngeal wall* minutely plicated;
horizontal bars usually broad and narrow alternately, their breadth never
exceeding half the length of the meshes; internal longitudinal bars
slender; papillæ above the connecting ducts erect, discoid, provided with
a supporting ridge in front and behind; no intermediate papillæ; meshes
square, each containing four or five stigmata. *Œsophageal aperture* on
dorsal side of pharynx, near its posterior end.

The *body* in all the specimens is much depressed, oblong in form,
with sloping and expanded edges, and attachment is affected by the
whole of the left side. The position of the oral and cloacal apertures
is indicated in fig. 7 (Pl. VI.), which represents the largest individual
of twice the natural size. The cloacal aperture varies slightly in

position, but it is always nearer the posterior than the anterior end of the body, between half and two-thirds of the way down the dorsal edge. Both apertures are small and inconspicuous; no ocelli were observed in the living animal.

The dimensions of the largest specimen are as follows :—

Maximum length	.	$1\frac{3}{8}$ inch
,, breadth	.	$\frac{3}{2}$,,
,, thickness	. . .	$\frac{1}{4}$,,

The *test* is firm and cartilaginous, though rather thin; it is not rough to the touch, but its surface is in reality studded with minute tubercles of bluntly conical form. They are so small that they cannot be readily observed when the test is immersed in alcohol, but when removed for a moment from the fluid, the presence of minute projections is detected by the broken reflection of light from its upper surface. A portion of the surface of the test is shown on Pl. VI., fig. 9, considerably magnified. A series of vertical sections through the test shows that the tubercles are quite solid, and that the culs-de-sac of the test-vessels have no connection with them. The greater part of the test is composed of huge "bladder-cells," the largest of which are as large as many of the tubercles on the surface; they are of spherical or polyhedral form. The superficial tubercles are entirely destitute of bladder-cells.

The body when deprived of the test is at least twice as long as broad in the majority of the specimens, but in one individual the proportion between the two dimensions is slightly less than this. The oral and cloacal siphons are both rather long and tubular, and the cloacal siphon is particularly so (Pl. VI, fig. 8).

From the œsophageal opening being situated near the posterior corner of the pharynx, the viscera extend to the posterior end of the body. The *stomach* is rounded in form and considerably wider than the intestine. The course of the intestine has been sufficiently indicated above.

The *ganglion* is remarkably elongated, being six times as long as broad; it extends from the level of the fifteenth to that of the twenty-first horizontal bar in the specimen shown in figs. 7 and 8.

The *epipharyngeal groove* in the same individual is a low furrow, not elevated behind, extending from the dorsal tubercle as far as the

ninth horizontal bar, but at the sixth bar its left lip suddenly thins
out and bends over the right lip, concealing it from view, and con-
tinuing posteriorly as the *dorsal lamina*. This structure has a very
characteristic form in this species (Pl. VI. fig. 10). It is moderately
deep, provided with a regular succession of transverse ribs on its
convex side and of well-marked teeth on its edge, the latter corresponding to the number of ribs. There are no intermediate pectinations
of its edge ; but the concave side instead of being smooth as is usually
the case in Ascidians, is profusely studded with stout papillæ, as
shown in the figure. There is a certain tendency of the papillæ to be
arranged in rows directed obliquely from the summit to the free edge
of the lamina, from before backwards ; but this general tendency is
frequently departed from. The dorsal lamina is continued for some
distance behind the œsophageal aperture.

Branchial apparatus.—The horizontal vessels are often of two
sizes,* broad and narrow, and these vessels alternate with one another
in position ; but the breadth of the larger vessels never exceeds half
the antero-posterior diameter of the meshes—usually it is considerably
less. The pharyngeal wall is minutely plicated. The internal longitudinal bars are slender in form. At their junctions with the
connecting ducts are situated blunt papillæ of characteristic shape ;
they are of an erect discoid form, with a semi-circular edge, compressed from before backwards, and provided with a supporting ridge
or buttress upon their anterior and posterior faces (fig. 11 *b*). Usually
the meshes are square, and intermediate papillæ quite absent ; but in
some parts of the pharynx transverse rows of meshes may frequently
be observed which are distinctly elongated in a longitudinal direction,
and in such regions minute intermediate papillæ may be detected upon
the internal longitudinal bars. The elongation of the meshes and
appearance of intermediate papillæ is preparatory to the formation of
a new series of horizontal vessels, in the manner which I have
described above in *Ascidia mollis*. There are four or five stigmata in
each mesh (fig. 11 *a*). There is no pharyngo-cloacal slit.

My largest specimen is mature, and minute white ova are present
in the oviduct.

* This distinction of size is much less apparent in mature than in young
individuals.

After much consideration I have arrived at the conclusion that the specimens whose structure I have just described represent the species *Ascidia depressa* of Alder, and that the Ascidians described under this name by Heller, Herdman, and Roule are distinct from it.

A reference to Alder's original account will show how perfectly in every point my specimens agree with his description, with the exception that I can make no statement as to the presence of red ocelli around the apertures. I did not observe these spots in the living animals; but on the other hand I paid no attention to the point, and probably overlooked their existence. In every other respect the correspondence is complete, and I may draw especial attention to the following details—the shape and colour; the expanded edge; the position and form of the apertures; the granulations (minute tubercles) on the upper surface; the absence of intermediate papillæ in the branchial sac (for it was Alder's habit to imply the absence of these structures when he made no direct reference to them); and the size.

If this be really so, it necessarily follows that Heller's specimen described under the same name is distinct. The structure of the test is alone sufficient to distinguish my specimens from his. The bladder-cells in the former are huge, of spherical or frequently polyhedral form, exactly as Heller has himself described and figured for his *Ascidia rudis* (l. c., p. 14, Taf. v, fig. 6); but for his *A. depressa* a very different condition was described by him (l. c., p. 15). Further, Heller's *A. depressa* was destitute of the superficial microscopic tubercles which are present in my specimens (and in Alder's) and which Heller himself also figured for another species (*A rudis*, l. c.).

Secondly, the specimens which Prof. Herdman has referred to this species differ from Alder's in possessing intermediate papillæ on the internal longitudinal bars; Alder would certainly have noticed the existence of such papillæ as Herdman has figured (l. c., *supra*, pl. xviii, fig. 4), if they had existed in his specimens. Prof. Herdman's specimens cannot belong to the same species as these from the Isle of Wight, because in the latter the internal longitudinal bars rarely show a trace of intermediate papillæ, except when the meshes have grown to a size when they are almost twice as long as broad; in Prof. Herdman's species these papillæ are normally present, and the meshes are elongated transversely. Further, the structure of the dorsal lamina

is very different in the two cases.   Prof. Herdman in the same paper
noticed the existence of tubercles on the dorsal lamina of *Ascidia
plebeia*, so that there is no reason to suppose that he overlooked them
in his *A. depressa*.

Lastly, M. Roule has described under the name *Ascidia depressa*, a
species which, while probably identical with Heller's, is undoubtedly
distinct from Alder's species. The mode of attachment, the shape of
the body, and the structure of the branchial sac are very different in
the two cases.   The species described by Heller and by Roule presents
a close affinity with *Ascidia mentula*, and still more, perhaps, with
Alder's (not Heller's) *Ascidia rudis ;* but there is nothing in Alder's
description of *A. depressa* to indicate a similar relationship for that
species, and my specimens are distinctly against it.

Ascidia depressa, as now re-described, is very closely related to
Traustedt's *Ascidia (Phallusia) pusilla* (Mitt. Zool. Stat. Neap., iv,
1883, p. 465, Taf. xxxiv, figs. 16, 17 ; Taf. xxxv, fig. 26).   The chief
points of difference are found in the different proportions of the length
to the breadth of the body, the length of the siphons, the breadth of
the largest horizontal vessels of the pharynx, the number of stigmata
in the meshes, the shape of the stomach, and especially the structure
of the dorsal lamina.   Some of these differences are trivial, and it is
impossible at present to say whether Traustedt's single specimen of *A.
pusilla* is, or is not, merely an abnormal individual of our species ; but
the constancy in the structure of the dorsal lamina in my specimens
is, when associated with the other peculiarities, a strong piece of
evidence in favour of the specific distinctness of the two types.

Ascidia depressa is also allied to *Ascidia marioni*, Roule, on account
of the close agreement between the two species in the following points—
the mode of fixation, the position of the apertures, the minute
tuberculation of the surface, the absence of intermediate papillæ, the
strong pectination of the dorsal lamina, the elongation and approximation
of the stigmata ; but the two species are of course quite distinct owing
to the important difference between them in the structure of the
subneural gland and its accessory organs.

I have already pointed out the curious resemblance between *Ascidia
depressa* and Heller's *Ascidia rudis* in the histological and superficial
structure of the test.   Since Heller's specimen agrees with Roule's

Ascidia marioni both in the position of the cloacal aperture and in the minute tuberculation of the surface, it is not improbable that the two are specifically identical ; but whether Heller's individual was rightly referred to Alder's species or not is very doubtful. Alder's *rudis* possessed " small, *distant* tubercles " on the test, and was "sometimes nearly smooth,"—a condition very different from that in Heller's specimen, as well as in Roule's *A. marioni*.

If, as I believe it will now be generally admitted, the forms described by Heller, Roule, and Herdman under the name *Ascidia depressa* can no longer lay claim to that title, it will be necessary to refer to them under new designations. I would propose for the Mediterranean species described by Roule the name *Ascidia Roulei*. To the variety *petricola* of this species, Heller's specimen almost certainly belongs. I believe that *Ascidia Roulei* is closely related to, if not identical with Alder's *Ascidia rudis :* but it is impossible to give a final decision upon this question until our British Ascidians have been collected and re-examined in greater detail.*

The form described by Herdman as *Ascidia depressa*, in the paper to which reference has been made above, appears to be distinct from *Ascidia Roulei*, although it is impossible, from the want of correspondence between the descriptions, to speak decisively. But Prof. Herdman has himself thrown doubt upon their identity in his recently published† Revised Classification of the Tunicata, so that a new name is, at least provisionally, desirable. I therefore propose for it the name *Ascidia Herdmani*.

The subjoined synonymic lists show briefly the conclusions to which I have been led by the study of this species from the Isle of Wight.

1. ASCIDIA DEPRESSA, *Alder*, 1848, loc. cit.

    = ASCIDIA DEPRESSA, *Garstang*, 1891 (the present paper).

   ?? = PHALLUSIA PUSILLA, *Traustedt*, 1883, loc. cit.

   non ASCIDIA DEPRESSA, *Heller*, 1875, loc. cit. (= *A. Roulei*, Garstang, 1891).

   nec   — —   *Herdman*, 1881, loc. cit. (= *A. Herdmani*, Garstang, 1891).

   nec   — —   *Roule*, 1884, loc. cit. (*A. Roulei*, Garstang, 1891).

* It is needless to say that we look forward with interest towards Prof. Herdman's promised re-description of some of Alder and Hancock's types.

† Journ. Linn. Soc. Zool., vol , xxiii, 1891, p. 594.

2. ASCIDIA RUDIS, *Alder*, 1863.   Ann. Mag. Nat Hist. (3), ii, p. 195.

    ? = ASCIDIA ROULEI, *Garstang*, 1891.

            = A. DEPRESSA, *Roule*, 1884 (non Alder, 1863, nec Herdman, 1881).

                [=VAR. PETRICOLA] A. DEPRESSA, *Heller*, 1875.

non   —   RUDIS, *Heller*, 1875.

       ? = A. MARIONI, *Roule*, 1884.

III.

Ascidia mentula, *O. F. Müller.*

ASCIDIA MENTULA, *Müller.*   Zoologia Danica, 1788, vol. i, pp. 6, 7, pl. viii, figs. 1—4.

    —   RUBROTINCTA, *Hancock.*   Ann. Mag. Nat. Hist., 1870.

    —   RUBICUNDA, *Hancock.*   Ibid., 1870.

    —   ROBUSTA, *Hancock.*   Ibid., 1870.

PHALLUSIA MENTULA, *Kupffer.*   Jahresb. d. Komm. z. Unters. d. deutsch. Meere in Kiel, Berlin, 1874, p. 209, pl. iv, fig. 1.

ASCIDIA, MENTULA, *Heller.*   Untersuchungen, 1875, loc. cit., pp. 2—13, pls. i—iv.

    —   RUBESCENS, *Heller.*   Ibid.

    —   LATA, *Herdman.*   Journ. Linn. Soc., xv, 1881.

PHALLUSIA MENTULA, *Traustedt.*   Die einfachen Ascidien, 1883, loc. cit., pp. 457—459.

ASCIDIA MENTULA, *Roule.*   Recherches, 1884, loc. cit.

Several large Ascidians, which I refer to this species, were found attached to the sides of a rock, situated far out in the Zostera bed off Nodes Point, on May 7th, at extreme low water, spring-tides having then almost reached their height. The following descriptions refer to two individuals which I brought away with me for more detailed examination ; they are given separately in order to indicate the degree of variation the more naturally.

A. *Body* oblong, elongated, attached by almost the whole of the left side. Dimensions—Length, 3 inches; breadth, 1⅓ inches; thickness, 1½ inch. An idea of its external appearance may be gained from the figure which Heller gives (l. c. pl. v, fig. 5) to represent a supposed specimen of Alder's *Ascidia rudis*, but the position of the cloacal aperture is different.

Test thin, hard, cartilaginous, greatly wrinkled in a longitudinal direction on the right side, almost entirely overgrown by small algæ, and extensive colonies of the Polyzoon *Alcyonidium mytili* and some Didemnids. Here and there on the right surface a few minute tubercles may be detected. *Oral aperture* on the right side, sub-terminal, not prominent, bounded by nine lips ; *cloacal aperture* on the right side near the dorsal edge, very slightly nearer the anterior than the posterior end of the body, bounded by six lips.

Upon removal of the test the rest of the body is seen to be of a yellowish colour, the musculature being of a rather deeper amber-colour. The *oral* and *cloacal siphons* are tubular but short. The oral siphon terminates in nine sub-triangular lips, which are rather prominent, with rounded apices and with a spoon-shaped concavity on their external surfaces. The edge of the siphon is bounded by a thin red line which is discontinuous towards the tips of most of the lips. A small red ocellus is found behind the red line between each pair of lobes, and the surface of the siphon is slightly sprinkled with red dots. The cloacal siphon terminates in six lobes, bounded similarly by a thin red line, but without ocelli. It is directed straight towards its external orifice.

Musculature coarse and strong, the fibres amber coloured.

Viscera disposed as usual in the species, the posterior border of the stomach being nearly ½ inch from the posterior end of the body ; the anterior wall of the intestine at its first bend is on a level with the ganglion ; the posterior wall at its second bend is on a level with the opening of the œsophagus into the pharynx.

Renal vesicles large, forming a soft yellowish coating over the stomach and intestine ; concretions showing as a small brown spot in each vesicle, when looked at with a lens, but resolving themselves in each case into a compact mass of several concretions, of different sizes and of a yellowish-brown colour, when examined under a low power of the microscope (cf. Roule).

Tentacles about thirty in number, short, of unequal sizes, irregularly arranged.

Prabranchial zone studded with microscopic papillæ arranged more or less regularly in longitudinal rows.

Dorsal tubercle longer than broad, presenting two apertures, one behind the other. The anterior is crescentic, with the horns produced and curved inwards ; the posterior is crescentic, with the left horn slightly produced and curved towards the mid-dorsal line, and with the right horn also curved round and produced a little beyond the mid-dorsal line (Pl. VI, fig. 12).

Epipharyngeal groove present for a short distance and then ceasing abruptly (fig. 12). The *dorsal lamina* is quite absent anteriorly, and does not appear until halfway between the position of the ganglion and the level of the pharyngo-cloacal slit, when it gradually rises up in the form of a narrow membrane and is continued to the posterior end of the pharynx. Dorsal lamina strongly ribbed transversely and minutely pectinated at the margin, the teeth corresponding to the ribs ; no intermediate pectinations ; concave side smooth.

A *pharyngo-cloacal slit** present on the right side of the dorsal

---

* I give this name to the curious aperture, so commonly found in the pharyngeal wall of *Ascidia mentula*, in which species it was first noticed by Kupffer (l. c.). It has been ingeniously suggested lately that it represents the persistent internal opening of the right primitive atrial canal, in spite of the fact that it is absent in the more primitive Ascidians, such as *Clavelina* and the *Distomidæ*. Now, as has been stated above (pp. 134 and 135), I have discovered this slit to be present in large individuals of two other species of Ascidians which are not closely allied to *Ascidia mentula* (*Ascidiella aspersa* and *Ascidia mollis*), although it does not exist in young specimens of those species. This fact is a sufficient disproof of the theory which gives to the slit the value of a phylogenetic remnant. My own theory is less attractive, but possibly more true. The slit is always situated opposite the cloacal orifice, and only occurs in large species (*Ascidia mentula* and its close allies *e.g.* of *Ascidia lata*, Herdman) and in large individuals of smaller species (*e g.* of *A. mollis* and *Ascidiella aspersa*). May it not be a special adaptation for the prevention of the over-accumulation of fæces in the cloacas of large Ascidians, where the ordinary methods of ejection are insufficient? Ascidians, being sessile animals, are especially liable to danger from such over-accumulation, as Giard long ago stated in the case of the *Didemnidæ* and *Polyclinidæ* (Arch. Zool. Exp., i, p. 520); and special means are adopted in various sections of the group to ward off the danger. For instance, as Maurice has well suggested, the cloacal languettes of the *Polyclinidæ* serve the definite function of keeping open the cloacal canals in colonies of that family (Arch. de Biol., viii, 1888, p. 213); while in the *Botryllidæ* the end is achieved only by the united efforts of the zooids in a cœnobium : they simultaneously and suddenly contract their bodies, and so drive a strong current of water through their peribranchial cavities into the common cloaca, ejecting the fæces with such violence, as Gaertner observed, "ut ingenti saltu oppositum faveæ marginem transiliant" (see Giard, loc. cit.).

In the large Ascidians under discussion, the presence of this big oval slit—it is frequently over a centimetre in length—directly opposite the cloacal cavity, will enable the animal, by a strong contraction of the muscular tunic, to drive lamina, directly opposite the cloacal aperture; slit, ¼ inch long and somewhat curved.

Dorsal tubercle horse-shoe shaped, midway between the slit and the dorsal tubercle.

Œsophageal opening high up in the pharynx, between the slit and the posterior third of the body. Behind it is a long smooth "post-buccal raphe" (see Heller's figure, l. c.).

Branchial apparatus.—Meshes elongated transversely; stout conical papillæ at the junctions, provided with supporting ridges in front and behind (fig. 13); intermediate papillæ equally long, but more slender than the primary papillæ; six or seven stigmata in a mesh; minute plications deep, the longitudinal furrows frequently bifurcating.

B.—The second individual differs from the one just described in the external form, and in the absence of any malformation of the dorsal tubercle and lamina; in other respects it is closely similar to the first specimen.

Body of a compressed pyriform shape, the narrow end anterior, attached by a circular area over the posterior half of the left side. Dimensions—Length, 2¼ inches; Maximum breadth across middle, 1⅜ inches; Thickness, ¾ inch.

Test very slightly furrowed, overgrown with algæ and Polyzoa.

Oral aperture terminal; *cloacal* on the dorsal edge, slightly nearer the posterior than the anterior end of the body.

Oral siphon with very short and obtuse lips; no red pigment upon either of the siphons.

Tentacles forty in number, considerably longer than in the preceding specimen, irregularly arranged.

a considerable body of water from the pharynx into the cloaca, and thus to effect the desired object more thoroughly than is possible when stigmata exist alone.

Kupffer has also recorded the existence of paired pharyngo-atrial slits, symmetrically placed in the posterior region of the pharynx, in *Ascidia conchilega* and *Ciona* [*canina*] *intestinalis*. The former species I have been unable to examine, but in *C. intestinalis* (preserved material) some individuals possess huge slits, through which the intestine conspicuously projects into the pharynx, while in other individuals no unusual apertures can be made out at all. (Cf. Traustedt, loc. cit., p. 453. Heller, loc. cit., ii, p, 118, seems merely to repeat Kupffer's statement. Roule, loc. cit., makes no reference to any exceptional openings.) I am inclined, therefore, to believe that in both these species Kupffer's apertures are accidental or artificial rather than natural.

K

Dorsal tubercle circular in shape; aperture horse-shoe shaped, the right horn curved inwards.

Epipharyngeal groove considerably longer, its lips gradually narrowing and becoming continuous with the dorsal lamina.

In all other respects this individual agrees with the former.

Both individuals are mature and have ova and spermatozoa in their generative ducts.

I believe that in point of size these specimens have undergone a considerable reduction since their capture. In the rough notes which I then made, I put down the length as "about 5 inches," while actual measurement now shows that the largest of the two brought away does not exceed 3 inches. Allowing for a possible degree of error in my original estimate of their size, there must still, I think, have taken place some contraction of their test and body in the four months during which they have been in alcohol. It is, I admit, unsafe to argue upon these grounds, for the larger ones may have been just those which I dissected at the time of capture and did not retain. I will, therefore, merely state that the size of some of the specimens which I found was fully 4 inches.

The *colour* of the individuals when alive was hardly different from that which these spirit specimens now exhibit. It is sufficient to say that there was an almost total absence of red pigment in their bodies, and what did exist was confined to the region of the siphons, particularly the oral siphon. The test-vessels, also, with their terminal dilatations, were destitute of red and of all conspicuous colouration.

The species *Ascidia mentula* has been described in greatest detail upon Mediterranean specimens, although it is widely distributed round all the coasts of Europe, and has been called the commonest of the British deep-water Ascidians. Off the south-western shores of England, however it is certainly not common within the 40 fathom line; I have only taken it once or twice there, and its place seems to be occupied by two other large Ascidians, *Phallusia mammillata* and a course variety of *Ascidiella aspersa*. Indeed, the fact that there is extant no anatomical description of British specimens referred to Müller's species, seems at first strange, if they are really so abundant. A comparison of my specimens with Müller's original description revealed some distinctions which at the outset seemed to be of some importance.

Both of Müller's specimens were brilliantly pigmented, the whole of the body within the test being of a bright crimson colour, except over the area occupied by the viscera on the left side, which was whitish, the intestine being of a livid green colour ("colorem luridum exhibens").

But in Traustedt's specimens from Naples the red pigment was found to be a very variable and unreliable characteristic; sometimes the stomach only was so coloured, sometimes this pigment was spread over the entire area of the branchial sac (as in Müller's specimens), whilst sometimes individuals were taken which were quite destitute of red colouration.

Roule, at Marseille, has observed that the test is almost always rose or red in colour, and he gives some beautiful figures in illustration of this condition, but he also admits a considerable degree of colour-variation in the species, which he attributes to local influences.

Heller's specimens from the Adriatic seem to have been much more subdued in colour than those from the neighbourhood of Naples and Marseilles. He describes the colour as "greenish or yellowish white, seldom brownish, the oral syphon usually edged with red (rothgesäumt);" further on he adds that the blood corpuscles are brownish. My specimens, therefore, approach Heller's very closely in this respect.

Now a perusal of Hancock's paper on *Several New Species of Simple Ascidians* (1870, l. c.) shows that he attached a considerable importance to distinctions of colour in his definitions of species, an importance which can no longer be admitted for *mentuloid* forms at any rate; and Roule has quite rightly, in my opinion, merged Hancock's *A. rubro-tincta* into the species *A. mentula*. *Ascidia rubicunda* of Hancock agrees perfectly with the typical *mentula* of Müller in its brilliant colouration, and I shall show below how unimportant is the only other character which distinguishes it from the general form of that species. *Ascidia robusta* of Hancock is distinguished from the specimens which I have described from the Isle of Wight by hardly any other point than the prolongation of the oral and cloacal siphons.

It may be observed that in all the *mentuloid* forms there is a distinct correlation between the position and extent of the area of attachment and the zone of the sea bed from which individuals have been taken.

The *Ascidia mentula* of authors is an inhabitant of the deeper waters, and is found attached usually to stones and shells by its base and a very little of the left side. Adhering in this way, it is obvious that it has an erect position upon the sea-bottom. Now the three species named above were distinguished by Hancock from *Ascidia mentula* partly on account of the mode of their attachment; *A. rubrotincta* adhered "by the middle portion of the side," *A. rubicunda* "by the whole side with imperfect marginal expansions," *A. robusta* "by the whole side, but [was] sometimes much distorted, and with adherent root-like prolongations."

These three "species" were all taken from between tide-marks, the first at Guernsey, the second at Tobermory (Isle of Mull), Portaferry (Strangford Lough), and Bertraghbuy Bay (Connemara), the third at Herm.

The Isle of Wight specimens also were attached by the whole or the greater part of the left side, and they also were taken from a rock at low water.

Now no one can have much attended to the conditions of existence in the littoral zone without having been impressed by the extent of the disturbing forces which littoral animals have to resist, if they are to survive in that locality. They are battered by the waves almost incessantly, and cannot exist without special means of defence. This defence in many groups is ensured by the development of strong adhesive or clinging organs, the prevalence of which among littoral animals shows, by a reversal of the argument, the extent of the disturbing forces that play around them.

Tunicates are essentially plastic creatures, for the structure and mode of development of their tests renders their external form easily modifiable. It would, therefore, be extremely improbable to find that the larvæ of *Ascidia mentula*, when carried by in-flowing tidal currents from deeper water into the littoral zone, would grow in quite the same way in one place as in the other. The incessant motion of the water would necessitate, and indeed frequently bring about, as growth proceeded, a larger area of attachment than would suffice to resist the comparatively feeble currents of deeper water.

The results of such a process would be (1) Hancock's *Ascidia rubicunda*, which is merely the red-coloured variety of *A. mentula* adapted to a littoral existence ; (2) my specimens from the Isle of Wight, which are merely the pale variety of *A. mentula* adapted to a littoral existence upon a comparatively smooth surface of rock ; (3) Hancock's *A. robusta*, which is a pale reddish variety modified in its mode of attachment by tidal influences, and in its general shape by the irregularity of the surrounding objects ('roots' of *Laminaria digitata*).

Even Müller a hundred years ago recognised the plasticity of form in his species, for, referring to the oral and cloacal apertures, he says : "Pro figura massae, quae ab adjacentibus corporibus determinatur, vel utraque lateralis, vel altera plerumque terminalis."

If it should be objected that the Mediterranean zoologists can supply little or no evidence of variability in the extent and mode of attachment in their specimens, the fact is rather in favour of my contention than against it ; for the causes to which the variation has been here attributed are absent in the Mediterranean, where the tidal oscillation, with its accompanying disturbance of the sea bottom, is so small that it may practically be neglected.

With regard to internal structure, the differences between the Isle of Wight specimens and those described by the Mediterranean zoologists are very slight and unimportant.

For a comparison of the descriptions of Mediterranean forms shows that variability is not confined to points of colour and external form. Traustedt gives the number of tentacles as from 78 to 85 in Neapolitan specimens, while Heller, who also examined the species in great detail, ascribes from 30 to 35 to Adriatic examples. There are 30 in one of mine, 40 in the other. Herdman's *Ascidia lata* (3¼ inches long ; one specimen) possessed from 16 to 20, and the species was defined upon the ground of this difference* and of a peculiarity in the aperture of the dorsal tubercle.

Take again the dorsal lamina. Heller unfortunately gives no details upon this point, but Traustedt and Roule agree that the lamina is strongly pectinated. In Roule's specimens the right face of the lamina is also provided with a few smaller "languettes." On

---

* Since the above was put in type, I have been enabled to examine some specimens of *A. mentula*, which were dredged in Loch Long and are now under Mr. Hoyle's charge in the Manchester Museum. The number of tentacles is so variable as to be only 18 in an individual 1¼ inches long, while it is nearly 40 in an individual 3 inches long.

the other hand, Hancock's *A. rubicunda*, Herdman's *A. lata* (from Loch Long), and my specimens agree in being merely minutely denticulated along the edge of the lamina.

It is true, therefore, that we have at last arrived at a point wherein some of the north Atlantic forms agree to differ from their Mediterranean relatives; but he would be rash who would distinguish the species upon this ground alone, in view of the numerous cross-resemblances in other respects.

The præbranchial zone is minutely tuberculated in my specimens just as in Traustedt's.

Altogether, therefore, there appears to be no sound reason why the numerous *mentuloid* forms mentioned in this paper should not be grouped together into one species and entitled *Ascidia mentula*. Some other "species" might even be added to the list. Heller's *A. rubescens* has rightly been included by Roule as a young individual of the species, and it is just possible that Herdman's *A. fusiformis* (1¾ inches long; three specimens) is merely a young condition also.

It is difficult to form an opinion upon Hancock's *A. plana* and *A. alderi;* but they appear to belong to this species also.

I cannot hope to have altogether avoided error in the course of this paper, but I have certainly endeavoured to do so; and I trust that, as an attempt to throw a little light upon some of our British Tunicates, my essay will not be without useful results.

Moreover, it would seem to be serviceable if a word or two should be said upon the desirability of keeping in mind the facts of variation, and of adopting some method by which the broad phenomena of variability within the limits of a species can be properly and systematically recorded.

It is now a truism that variation does not only consist in the manifestation of irregular abnormalities. The commonest Anemone of our sea coasts, *Actinia equina*, Linn., sufficiently testifies to the existence of a fixity and a stability even in variation. Yet it would be a strange misconception of the species-idea that would lead anyone to specifically separate the more constantly varieties of *Actinia equina* or of *Cylista undata* from one another simply on the ground of that constancy.

The _____ _____ of taxonomy is a great boon to the investigator in biology, but it becomes a burden when it is applied with random pen to every little group of forms, distinguished though they may be, under their particular conditions, by the constant possession of some minute peculiarity. Minute and constant peculiarities are of the greatest interest and importance, and nothing could be, for some time to come, of _____ value to the student of organic evolution than their careful recognition and classification, involving also a similar record of the bionomical conditions under which those peculiarities are found to be manifested.

But there is no reason why the specific name should be bestowed upon these minutely isolated groups. They had much rather have a nomenclature of their own within the limits of the species embracing them ; and that such a nomenclature can be adopted with success is sufficiently established by a perusal of Mr. Gosse's admirable monograph of the British Actinians,—to go no further.

I will conclude with an attempt, by way of illustration, to record what seem to be the main outlines of variability in the species which has just been discussed.

ASCIDIA MENTULA, *O. F. Muller.*

Var. 1.—RUBERRIMA. Body-walls beneath the test of a brilliant red or rose-colour ; tentacles (always ?) numerous (60 to 80).

Form α.—*Erecta.* Area of attachment small, usually posterior and basal ; infra-littoral.

Distribution.—Off the south coast of Norway ; Mediterranean, off Marseille and Naples, rare in Adriatic (=*A. rubescens,* Heller).

Form β.—*Depressa.* Area of attachment large, extending over the whole or the greater part of the left side ; littoral.

Distribution.—West coast of Scotland, west and north-east coasts of Ireland (=*A. Rubicunda,* Hancock).

Var. 2.—RUBROTINCTA. Body-walls tinged with reddish flesh-colour.

Form α.—*Erecta.* Attached as described above ; infra-littoral. Naples, Marseille, British seas ?

Form 3.—*Depressa.*   Attached as described above;
littoral.
        Channel Isles (=*A. rubrotincta* and
        A. robusta, Hancock).

Var. 3.—FLAVA.   Body-walls yellowish, with little or no trace of red;
tentacles rarely exceeding 40 in number.

Form 1.—*Erecta.*   As above; infra-littoral.
        Adriatic.   [West coast of Scotland
        (=*A. lata*, Herdman; but the colour of
        this race is only known from spirit
        specimens).]

Form 3.—*Depressa.*   As above; littoral.
Isle of Wight.

DESCRIPTION OF PLATES V. AND VI.,

Illustrating Mr. W. Garstang's paper "On some Ascidians from the
Isle of Wight."

N.B.—All the figures were drawn from preserved material.

PLATE V.

FIG. 1.—*Ascidia mollis*, Ald. and Hanc.   The largest normal individual,
nat. size.

FIG. 2.—*A. mollis.*   Culs-de-sac of the test vessels, magnified.

FIG. 3.—*A. mollis.*   Another individual of smaller size, as seen after removal
of the test; twice the natural size.

a. = View of the right side, showing the musculature.

b. = View of the left side, showing the disposition of stomach and
    intestine.

FIG. 4.   *A. mollis.*   Portion of the pharyngeal wall of the same individual;
much enlarged. Zeiss, Obj. A. Oc. 2, Cam. luc.   The dark portions represent
the longitudinal furrows, the light portions the elevations which are caused
by the "minute plication" of the wall.

c.d. = Connecting ducts between the horizontal and the internal longi-
    tudinal vessels.

h.v. = Horizontal vessels, forming three complete series and a rudimentary
    fourth.

i.l.b. = Internal longitudinal bars or vessels.

i.p. = Intermediate papillæ.

p. = Papillæ on the int. long. bars above the connecting ducts.

     The large abnormal individual, nat. size.

     above the dorsal surface. The left side consists entirely of the area of attachment; the right side forms an elevated ridge. The inconspicuous slit-like oral and cloacal apertures are indicated.

  b. = View of the opposite surface.

FIG. 6.—*A. mollis.* The same with the test removed, in the same position as in fig. 5 *a.* Nat. size.

  an. = Anus.
  c.s. = Cloacal siphon.
  gn. = Ganglion.
  int. = Intestine—the descending portion.
  œs. = Œsophagus.
  o.s. = Oral siphon.
  ov. = Oviduct.
  pc = Pericardium.
  st. = Stomach, covered with renal vesicles.
  v.d. = Vas deferens.

PLATE VI.

FIG. 7.—*Ascidia depressa,* Alder. The largest individual, twice the natural size.

FIG. 8—*A. depressa.* The same, with the test removed, viewed from the left side, showing the course of the viscera, and the rather elongated siphons.

FIG. 9.—*A. depressa.* A portion of the test, magnified, showing the papillæ on its surface.

FIG. 10.—*A. depressa.* A portion of the dorsal lamina, magnified, showing the marginal teeth (*m.t.*) and the lateral papillæ which project from its concave surface. Camera lucida.

FIG. 11 *a.*—*A. depressa.* A portion of the pharyngeal wall, magnified. Camera lucida.

  h.v. = Horizontal vessels.
  i.l.b = Internal longitudinal bars or vessels.
  p. = Papillæ above the connecting ducts.
  r.i.p. = Extremely rudimentary intermediate papillæ, here and there present where the meshes are elongated.

FIG. 11 *b.*—*A depressa.* An enlarged view of the junction between an internal longitudinal bar (*i.l.b.*) and a horizontal vessel (*h.v.*) showing the form of the disc-shaped papilla (*p.*), with its anterior (*a.b.*) and posterior buttresses.

Fig. 12.—*Ascidia mentula*, O. F. Müller. The peritubercular area in the individual A., showing the double aperture of the dorsal tubercle.

ep.gr. = Epibranchial groove.

p.gr. = Pericoronal groove.

p.z. = Præbranchial zone, studded with minute papillæ.

Fig. 13.—*A. mentula*. Portion of an internal longitudinal bar (*i.l.b.*), seen obliquely from the side, showing the form of the papillæ on its surface; magnified. Camera lucida.

c.d. = Connecting duct.

h.v. = Horizontal vessel.

i.p. = Intermediate papillæ.

p. = Papillæ.

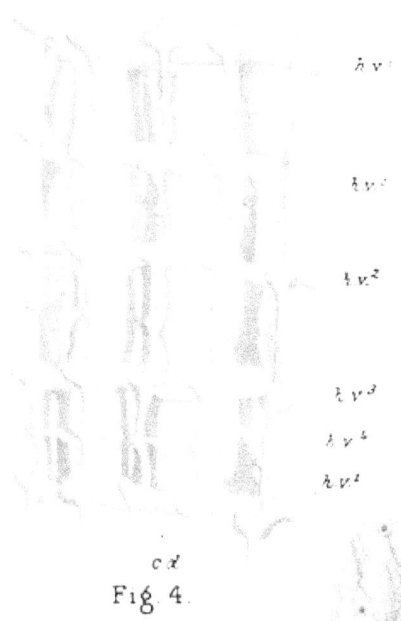

h.v¹

h.v¹

h.v²

h.v³
h.v⁴
h.v⁴

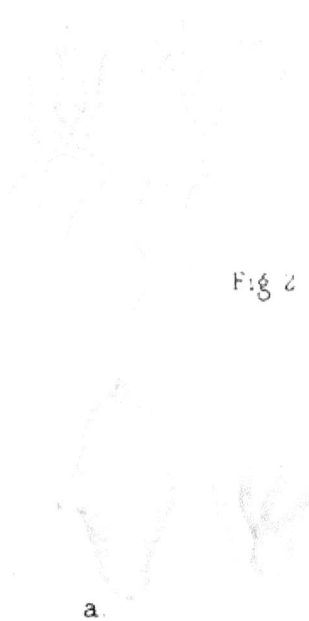

Fig. 2.

c.d

Fig. 4.

a.

Fig. 3.

rad.ac

anterior

Fig. 1

anterior

cloacal ap

os

g'.

right

int

e.s

v.d
cv
st
pc.

œs

right

posterior

Fig. 5. a.

Fig. 6

posterior

Fig. 5. b

Fig 7

Fig 8.

Fig 9

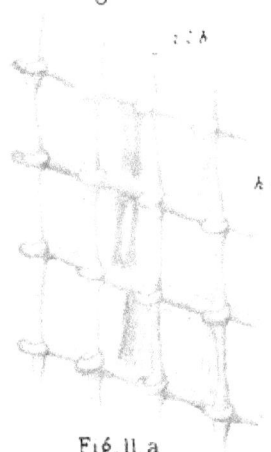

Fig. 11. a

Fig. 11. b

Fig 10

Fig. 13

Fig. 12.

From the ANNALS AND MAGAZINE OF NATURAL HISTORY *for May, 1892.*

OBSERVATIONS ON TWO RARE BRITISH NUDIBRANCHS,

(Lomanotus genei *Verany, and* Hancockia eudactylota, *Gosse*).

By F. W. GAMBLE, B.Sc., *Assistant to the Beyer Professor of Zoology, Owens College, Manchester.*

Plate VII.

While working last summer at the Plymouth Laboratory of the Marine Biological Association I obtained a single specimen of each of these species during successive weeks from the same part of Plymouth Sound. Finding that my *Lomanotus* possessed certain peculiarities of which I could find no adequate description or figures, and that *Hancockia* had only been taken on one previous occasion on the British coasts (by Mr. A. R. Hunt, in Tor Bay, 1877), I observed and drew the living animals with the following results.

Lomanotus genei, Verany. (Pl. VII. figs. 1 and 2).

Specimens referable to this species have been taken from time to time on our coasts. Mr. Garstang, in his recent report*, has collected these cases and added a number which have occurred at Plymouth. The following description of my own specimen agrees closely in certain points, such as size, colour, and general structure, with that of his two dark individuals †.

Length half an inch.

---

* "Complete List of Plymouth Opisthobranchs," Journ. Mar. Biol. Assoc. (n. s.) i. no. 1.

† "Report on Nudibranchs of Plymouth Sound," Journ. Mar. Biol. Assoc. I. ii. 1889, p. 187.

Colour dark brown, with irregular yellowish spots; the papillæ each
with a dark band below a white tip. The general tint agreed closely
with that of the *Fucus* on which I found it after being dredged, and
upon which it rested in captivity.

Oral veil with two prominent processes on each side, the outer ones
being the larger. Rhinophores retractile within calyx-like sheaths,
clavate, laminated at the base, with smooth truncate tips. Sheath-
margins produced into five papillæ of very definite shape when fully
expanded. These papillæ, like those of the oral veil and pleuropodium, are capable of contraction and dilatation. Pleuropodium
consisting of four well-marked lobes on each side. The centre of each
lobe is dorsal and close to the middle line. It is marked by the large
dorsal papilla. The sides of the lobe extend anteriorly and posteriorly
in a ventral direction, enclosing an area concave outwards, and bearing
papillæ. Posteriorly the lobes become slightly irregular and meet on
the dorsal surface. Foot slender, produced anteriorly into recurved
processes. Genital aperture beneath and slightly in front of the first
large dorsal papilla of the right side. Anus beneath the second.

My attention was first drawn to the characteristic form and changes
of shape assumed by the dorsal papillæ. These changes consisted of
contraction from an extended definite spiked shape to a more or less
bulbous triangular one. So far as I am aware none of the terms used by
previous authors on this subject do justice to the form of the extended
pleuropodial papilla. The interest of the matter is increased by the
fact that the tips of the "calyx-sheath" have the same power of
contractility, and that their extended form agrees with that of the
dorsal papillæ. The velar processes also when extended are of a very
definite shape (see figs. 1 and 2).

On gently touching the centre of the right side of the animal with
a clean sable brush three events occurred almost simultaneously; the
rhinophores previously expanded were sharply retracted within their
sheaths; the velar processes were extended; and the dorsal papillæ of
the right side, especially those near the point of the brush, were
erected from a previously oblique position, the large papillæ markedly
directing their whitish tips towards the brush. The effect might be
almost said to be "bristling." The papillæ of the left side were only
feebly affected. On repeating the experiment at different points I

found that when the stimulus is applied just behind the rhinophoral sheath the large postero-external sheath-papilla directed its tip obliquely backwards towards the point of attack, the first primary pleuropodial papilla directing its tip forwards. Several times I observed a single fully-expanded papilla move independently in an oblique plane from an anteriorly directed position to a posteriorly directed one. The "erection" and movement of the papillæ is brought about in the same way by natural stimuli. These movements led me to suspect the presence of cnidocysts. In spite, however, of the examination of the living animal and of sections of young specimens $\frac{2}{3}$ inch long (for the use of which, together with help in many ways, I am indebted to my friend Mr. Walter Garstang), I have hitherto been unsuccessful; indeed Bergh,[*] in his description of the genus, has stated "cnidocystæ nullæ" as a diagnostic character.

On some occasions I observed the peculiar lashing movements of the whole body already noticed by Mr. Garstang.[†] Thus, on pushing the animal laterally with a brush until its foothold gave way, it bent upon itself and executed a series of very vigorous S-shaped movements from side to side, the ventral surface of the foot being kept at about the same position on the surface of the water, while the rest of the body was inverted. On another occasion it voluntarily loosened its hold of the side of the glass vessel and progressed slightly by means of these contractions. Again, after floating foot upwards for some time, it would wriggle to the bottom and immediately gain a footing.

My specimen was quiet during the day. In the morning I found that it had crawled out of the dish where it had been placed overnight. This was done constantly. During the three weeks that I kept my specimen no spawn was deposited; hence probably it was immature.

As regards the significance of these observations. Continual changes of form in the pleuropodial papillæ during life have been noticed by Dr. Norman in his species, *L. Hancocki*.[‡] The complete similarity, however, both in characteristic form and power of co-

---

[*] "Die Cladohepatischen Nudibranchien," Zool. Jahrbücher, Bd. v. (1890).

[†] "First Report on Nudibranchs of Plymouth Sound," Journ. Mar. Biol. Assoc. (n. s.) I. ii. 1889, p. 189.

[‡] Norman, Ann. & Mag. Nat. Hist. 1877, xx. p. 518.

ordinative movement possessed by these papillæ in common with those of the "calyx-sheath" apparently escaped him, and is an additional argument in favour of the view advanced by Mr. Garstang,[*] that such sheaths contain a "pleuropodial" element."

Hancockia eudactylota, Gosse   (Pl. VII. Fig. 3.)

A specimen of this species was dredged last summer (1891) on *Delesseria* in Plymouth Sound, as I have already recorded.[†]   Mr. Hunt, the original discoverer of this form, dredged the only previous British specimen on the same Alga in Tor Bay in 1877.   This was described by Mr. Gosse[‡] under the name *Hancockia eudactylota*.   In 1886 Prof. Trinchese, apparently in ignorance of Gosse's paper, described ("Ricerche Anatomiche sul Genere *Govia*,"[§] 1886), four specimens dredged near Naples, defining them as two species of a new genus, *Govia rubra* and *G. viridis*.   Although the internal anatomy of *Hancockia* is unknown, it seems probable that the genera *Govia* and *Hancockia* will be united, as indeed has been done by Dr. Norman in his "Revision" (this Journal, Vol. vi., 1890, pp. 79, 80).   Carus ('Prodromus Faunæ Mediterraneæ,' vol. ii. pt. 1, p. 208) writes the genus *Govia*, Trinch., adding in brackets (*Hancockia*, Gosse).

The Plymouth specimen was about a quarter of an inch in length when expanded.   This is only half the length of Mr. Hunt's specimen. Colour a purplish-rose, very similar to the *Delesseria* on which it lived.   Too much stress should not be laid on this point, however, since Mr. Hunt's example, although apparently found on the same weed,[‖] was olive in colour.   The mid-dorsal and lateral lines of the upper surface darker.   The epidermis of the upper surface is of a bluish-green hue, as Gosse has already noticed (*loc. cit.* p. 317).   On the sheaths of the rhinophores are scattered bluish-white spots; semilunar markings of the same kind occurred at the base of the

* "Complete List of Opisthobranchs at Plymouth," Journ. Mar. Biol. Assoc. (n. s.) i. No. 1, p. 130.

† "The Occurrence of *Hancockia* at Plymouth," *ibid.* (n. s.) vol. ii., No. 2, p. 193.

‡ Ann. and Mag. Nat. Hist., ser. 4, vol. xx., 1877, p. 316.

§ Mem. della R. Acc. delle Sc. dell' Instituto di Bologna, ser. 5, vol. vii.

‖ Gosse, *loc. cit.* p. 316, note.

pleuropodial lobes (compare Trinchese's figure of *Govia rubra*). Body widest behind the head, gradually tapering posteriorly. Head with an oral veil bearing four papillæ on each side, the second anterior one being the largest. These papillæ were constantly changing their shape during life, as Gosse and Trinchese have recorded. Rhinophoral sheaths erect, cylindrical, the margin subdivided into about ten rounded projections. This agrees closely with the figure and description of the sheaths of *Govia viridis*. Those of *G. rubra*, on the other hand, have plain margins. Rhinophores with a rounded, bulbous, laminated base, terminating above in a smooth columnar tip. Pleuropodium produced into four lobes on the right and five on the left, the fifth being rudimentary. The first pair of lobes are opposite, the rest gradually becoming alternate, as in Trinchese's figure of *Govia rubra*. Each lobe is concave externally and is composed of seven papillæ, one being dorsal and median, three anterior, and three posterior. The foot is rounded anteriorly, posteriorly it ends in a slightly bifid tail, as in *Govia* (Trinchese, *loc. cit.* p. 183 and my fig. 1). The anal papilla very small, cylindrical, situated half way between the first and second lobes of the right side. Genital opening near dorsal surface between the rhinophore and the first dorsal lobe of the right side.

In the appended table I have compared the different specimens of *Hancockia* and *Govia*. Although they all agree in main points, no two individuals do so in detail.

	Name.	Length.	Colour.	Characters of Velar Papillæ.	Margin of Rhinophoral Sheath.	Pleuropodium.		
						Characters.	Number of Lobes.	
							Right side.	Left side.
1.	Govia rubra, Tr.	10 mm.	Purplish-rose. Brightest on rhinophoral sheaths, dorsal and velar papillæ. Dorsal surface with irregular white spots.	5 on each side; 3 large lateral, 1 small anterior and posterior.	Plain.	Lobes of 7-8 papillæ.	5; the 5th most posterior.	5
2.	G. rubra, Tr.	13 mm.	Colour as in No. 1.	4 on each side.	Plain.	Lobes of 7-8 papillæ.	5	6; the 6th most posterior.
3.	G. rubra, Tr.	Not given.	Bright chestnut. White spots forming a line marking the position of pleuropodial ridge.	5 on each side.	Plain.		Number not stated.	
4.	G. viridis, Tr.	14 mm.	"Verde pistacchio."	4 on each side.	Divided into about 10-12 rounded projections.		3	4
5.	Hancockia eudactylota, Gosse. Mr. Hunt's specimen.	12·5 mm.	Olive. Gosse figures whitish spots marking the pleuropodial ridge, becoming more irregular posteriorly.	3-4 on each side alternating with as many small ones (Gosse, loc. cit. p. 317).	Furnished with sub-conical points along its edge.	Lobes of 11 irregularly notched leaflets.	3	3
6.	H. eudactylota. Plymouth specimen.	7 mm.	Purplish-rose. White spots at base of pleuropodial lobes as in Govia.	4 on each side.	Divided into about 10 rounded projections.	Lobes of 7 papillæ.	4	3; the 5th most posterior.

From this table it would appear that *Govia rubra*, Tr., differs specifically from *G. viridis*, which may hereafter be united with *Hancockia eudactylota*, Gosse.

Our knowledge of the internal anatomy of these forms is limited to the preliminary paper by Prof. Trinchese before referred to. The cutting-edge of the jaw is short and armed with a single series of 15-16 teeth, the first two or three of which are simple, the rest set with extremely fine tubercles. Radula triseriate ; the teeth of the median row with lateral denticles ; the lateral teeth broad, unarmed ("quasi omnino illi Galvinarum similis," Bergh*). Salivary glands large. Liver diffuse, with anterior and posterior branches, the latter supplying the dorsal papillæ. The nervous system similar to that of Eolidiidæ. Eyes well developed. Otocysts with a single otolith. Penis unarmed. The spermatozoa similar to those of Eolidiidæ. *Hancockia* appears to be mature when about half an inch in length. Trinchese describes ripe generative products at this stage, and Gosse has figured and described the spawn deposited by a specimen of this size. The ribbon was in the form of two complete figure-of-eight coils, the ova being irregularly scattered. My specimen was only a quarter of an inch long, and during the fortnight that I kept it no spawn was shed.

I stimulated *Hancockia* to see if the dorsal papillæ would respond, as they do in *Lomanotus* ; no effect, however, followed. The presence of cnidocysts in the genus described by Trinchese as occurring at the tips of the pleuropodial lobes (*loc. cit.* pp. 186, 189, and plate, figs. 8 and 14) makes its behaviour contrast still more with that of *Lomanotus*.

While gliding over the bottom of the vessel in which it lived it would sometimes stop, raise the anterior part of the body, and, with the velar tentacles and the rhinopores well expanded, it would sway from side to side. In a short time the action ceased and the animal went straight to the *Delesseria* on which it lived. Unfortunately I made no experiments to ascertain whether *Hancockia* responds to shadows as stimuli. The large eyes noted by Trinchese would be in favour of such reaction. *Hermæa bifida*, which lives on *Delesseria*, and certain Eolids have been shown by Mr. Garstang to respond.†

As regards the systematic position of *Hancockia*. Gosse placed it in the Tritoniidæ ; Trinchese, Bergh, Norman‡, and Carus place it

* "Die Cladohepatischen Nudibranchien," Zool. Jahrb. v. p. 53.

† Garstang. "Complete List of Plymouth Opisthobranchs," Jour. Mar. Biol. Assoc. (n. s.) i. no. 4, p. 423.

‡ "Revision of British Mollusca," Ann. & Mag. N. H. vol. vi. 1890, p. 79.

L

in the Dotonidæ; Bergh, however, adding: "Bei der Formulirung
der Charaktere der Dotoniden ist auf die Hancockien oder Govien
keine Rücksicht genommen, weil die Stellung dieser merkwürdigen,
gleichsam mehrere Familien verbindenden Gattung, bei der bisherigen
nur vorläufigen Untersuchung Trinchese's noch ganz unsicher ist." I
will only allude here to one view implied rather than expressed by
Mr. Garstang.* He compared a lobe of the pleuropodium of *Hancockia*
with one of the four arcuate lobes of the "raised curtain" forming
the pleuropodium in *Lomanotus*. The side view which I give of the
latter genus shows that the lobes are distinct and that the breaks
occur between the segments having the large dorsal papillæ as their
centres (Pl. VII. fig. 2).

EXPLANATION OF PLATE VII.

Fig. 1. Plymouth specimen of *Lomanotus genei*, Ver., seen from the dorsal
     surface. × 6. The papillæ are extended.

Fig. 2. The same, from the right side. × 6. Papillæ about ½ expanded.
     a, genital papilla; *b*, anal papilla. These were inserted from the
     preserved specimen.

Fig. 3. Plymouth specimen of *Hancockia eudactylota*, Gosse, from dorsal surface.
     × 14. In this view only three papillæ of each pleuropodial lobe are
     visible.

* *Ibid*, p. 429.

Plate VII.

Fig. 2

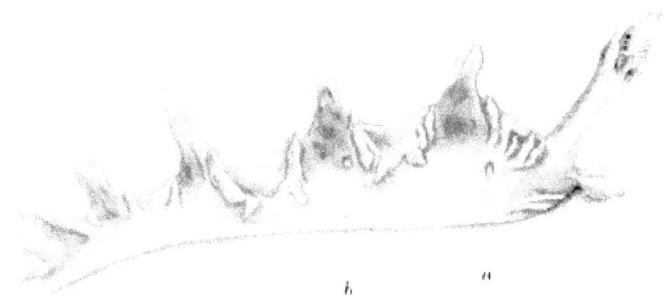

b   *a*

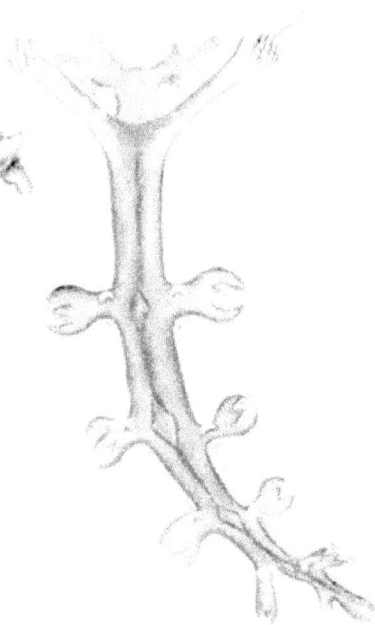

Fig. 1   Fig 3

1 & 2. *Lemanotus genei Ver.*   3. *Hancockia eudactylota Auasv.*

Reprinted from the QUARTERLY JOURNAL OF MICROSCOPICAL SCIENCE.

ON THE DEVELOPMENT OF THE OPTIC NERVE OF VERTEBRATES, AND THE CHOROIDAL FISSURE OF EMBRYONIC LIFE.

By RICHARD ASSHETON, M.A., *Demonstrator of Zoology in the Owens College.*

———

With Plates VIII. and IX.

———

In the 'Anatomische Anzeiger' of 28th March, 1891, Froriep described, in a somewhat brief manner, the way in which the fibres of the optic nerve are developed in the embryo of *Torpedo ocellata.* The paper contains twelve outline figures, which satisfactorily indicate that in the case of *Torpedo ocellata* at least certain fibres of the optic nerve arise from nerve-cells or neuroblasts situated in that portion of the optic cup which will form the retina at a later stage of development.

Herr Froriep is, I believe, the first who has published any figures in support of what is not altogether a new idea. In a foot-note to the above-mentioned paper Froriep remarks that Keibel had previously described the centralwards growth of the fibres of the optic nerve from the retina in reptilian embryos.

The original statement of Keibel's was in the form of a communication to a meeting of the Naturwissenschaftlichmedicinischer Verein in Strasburg, and, as I understand from a letter from Herr Keibel himself, there is not a fuller published account than that which appears in the 'Deutsche medicinische Wochenschrift,' 7th February, 1889, p. 115, which is as follows:—" Ueber die Entwickelung des Sehnerven.—Bei allen Amnioten bilden sich die Nervenfasern in dem unteren, mit der Netzhaut-anlage in Verbindung stehenden Theil des Augenblasenstieles ; eine secundäre Lösung der oberen Wand vom Pigment-Blatt der Retina und eine Verschmelzung derselben mit der eigentlichen Retina anlage findet nicht statt.	Präparate von Reptilienembryonen (Lacerta muralis und Tropidonotus natrix), welche mit Boraxcarmin gefärbt und mit Picrinsäure nachbehandelt sind, zeigen ferner, dass, bei diesen Thiere wenigstens, die ersten Schnervenfasern von der Peripherie centralwärts wachsen.	Es ist demnach höchstens wahrscheinlich, dass sie aus der Retina-anlage hervorwachsen, wenn ihr Ursprung dort auch nicht direkt nachgewiesen werden konnte."

The suggestion that the fibres of the optic nerve either develop from cells within the retina and grow towards the brain, or that they develop in the brain, and grow towards the retina, and are not formed by the transformation of the cells of the optic stalk, was first discussed as a matter of theory by His in the year 1868.

His (10) at that time maintained, as he has so ably demonstrated in a more recent work (11), that nerve-fibres are outgrowths from nerve cells ; and holding that there are no nerve-cells along the optic stalk, he suggested that the fibres of the optic nerve must grow from cells probably in the brain along the optic stalk to the retina.

With him in the main agreed W. Müller (20), Mihalkovics, and Kölliker, though they differed as to the direction of the growth of the fibres.

Certain other embryologists, however, were not convinced by what was little more than suggestion ; among them was Balfour (1, p. 493), who wrote, "There does not seem to me to be any ground for doubting (as has been done by His and Kölliker) that the fibres of the optic nerve are derived from a differentiation of the epithelial cells of which the nerve is at first formed ;" and Balfour's opinion seems to have been held until the present time by nearly every English writer on the subject.

There are apparently still those who hesitate to accept the more recent views as to the origin of the optic nerve.	As examples I will quote passages from the last editions of two widely read educational works, namely, 'A Text-book of Physiology,' by Foster, and 'The Frog,' by Marshall.

On p. 1141 of the former work Professor Foster writes : "The cup becomes what we may speak of broadly as the retina, and we may call it the optic or retinal cup; the solid stalk becomes the optic nerve." And again, on page 1142, " At the time when the epithelial cells of the stalk of the retinal cup are developed into the fibres of the optic nerve these become connected with the elements of the inner or retinal wall of the cup; they pierce the outer wall of pigment epithelium, making no connections with the cells of that outer wall."

On p. 131 of the latter work Professor Marshall writes : " The optic vesicles have already been described as arising at a very early period as lateral outgrowths from the fore-brain ; these soon become constricted at their necks, so as to be connected with the brain by narrow stalks, which ultimately become the optic nerves."

In view of this difference of opinion, which still exists concerning the origin of the fibres of the optic nerve, I think an account of a research made some months ago upon this subject in the frog and chick may be of some interest, especially as certain structures connected with the development of the eye and other parts of the nervous system may be more easily understood by appreciation of the fact that the optic nerve and optic stalk are two entirely separate structures.

Relation of the Optic Stalk to the Optic Nerve.

The two views held at present are—1. The optic nerve is formed by the differentiation of the cells of the optic stalk into nerve-fibres, which subsequently lose connection with the inner wall of the optic cup, and, piercing the outer wall, make connection with the outer face thereof.

2. The optic nerve is formed by the growth of nerve fibres either from the retina (outer wall of the optic cup) or from the brain, along the optic stalk, but outside it and unconnected with it.

The first of these views has been held by Balfour (1), Haddon (8), Foster (6), Marshall (19), &c. ; while the second view has been held by His (10, 12), Müller (20), Kölliker (14), Hertwig (9), Orr (21), and more recently supported by Keibel (13), Froriep (7), and Cajal (3).

In the new edition of 'Quain's Anatomy' Professor Schäfer seems to be uncertain which view to take. He writes on p. 79, " The optic nerves take origin as hollow outgrowths of the brain, which afterwards become solid, while nerve-fibres become developed in their walls. Their mode of origin will be further treated of in connection with the

development of the eye." On p. 85 Professor Schäfer gives what
appears to be an abstract from O. Hertwig's 'Lehrbuch' (9), pp. 402
—406, in which Hertwig gives His's more recent opinions on the
subject.

His (10) originally considered it probable that the nerve-fibres
arose within the brain and travelled towards the retina, but recently
(12) has changed his views, and considers it more likely that they
arise from neuroblasts within the retina and grow centralwards.

Thus it seems that the tendency of writers of this country is to
adhere to the older view in spite of strong evidence in favour of the
newer advanced by foreign authors.

Moreover, if we accept the theory that all nerve-fibres are outgrowths from nerve-cells, we have the advantage of knowledge of a
peculiarly fascinating nature, and the comprehension of the structure
and development of the central nervous system is rendered clearer to
both teacher and pupil.

The beautiful works of His, Cajal, and others—but more especially
of His—having once taught us what to look for, it is an easy matter
with ordinary care in the preparation of specimens and with thin
sections, to find the early commencement of nerve-fibre tracts, arising
as they do from definite groups of neuroblasts. I do not doubt that
the conclusions of His are correct, namely, that the fibres of the dorsal
roots of the spinal nerves and the sensory fibres of the cranial nerves
arise as processes from neuroblasts of the spinal ganglia in the one
case, of the cranial ganglia in the other case, and grow inwards to the
central nervous system.

Similarly also the sensory fibres of the sense-organs may be expected
to grow inwards from the sensory epithelium of the sense-organ to the
central nervous system.

In no sense-organ can the outgrowths of the fibres to the central
nervous system be more easily traced than in the case of the fibres
arising in connection with the optic organ; as the distance between
the place of origin of the fibres (in the retina) and the final destination
of the fibres (the brain) is relatively greater than in such cases as the
distance between the olfactory epithelium and brain, or between
ganglia of cranial and spinal nerves and the neural tube, in which
cases it is exceedingly difficult to trace the centralwards growth of the

nerve processes into the brain, though neuroblasts with the nerve processes directed towards the brain may be easily found.

I have paid special attention to the development of the nerve processes in the frog, and so shall describe the development of the optic nerve in that animal at some length.

Fate of the Optic Stalk.

I wish first to draw attention to the figs. 1, 2, 3, 8, 9, 10, 11, of Pl. VIII, which are camera drawings of sections which seem to me to prove conclusively that the optic nerve is not developed from the optic stalk ; that is to say, the nerve-fibres of the optic nerve do not arise by a transformation of the cells of the optic stalk into nerve-fibres.

During the earliest stages of the folding off of the optic vesicles the walls of the stalk are more than one cell thick ; but by the time the vesicles are definitely formed, and the outer wall has apparently begun to be pushed inwards upon the inner wall, the walls of the optic stalk are not more than one cell in thickness, and never become any thicker. The same statement may be made about the inner wall of the cup itself. From the time that the optic vesicles first form—that is, during the folding up of the neural plate—till after the fibres of the optic nerve have appeared, the optic stalk is hollow from end to end. With the folding off of the optic vesicles the optic stalk diminishes in diameter, and consequently the lumen also diminishes.

When the fibres first appear (in tadpoles of $6\frac{1}{2}$—$7\frac{1}{2}$ mm. in length) along the outer and ventral border of that part of the stalk nearest the optic cup, the lumen is still continuous throughout (figs. 6 and 7), but the greater part of the cavity between the outer and inner walls of the optic cup has become obliterated by the approximation of those two walls (fig. 12). The lumen of the optic stalk is first obliterated in tadpoles of about 10—11 mm. at the point at which the optic stalk and nerve pierce the dense tissue now forming at the side of the brain. At this point the optic stalk becomes squeezed, the lumen obliterated. Close to the brain the lumen persists for a long time, and is not entirely obliterated until the tadpole has attained a length of about 40 mm.

Fig. 1 is a horizontal section of the eye of a tadpole 10 mm. long, in which the fibres of the optic nerve can be seen along the whole

length of the optic stalk, and for a short distance into the brain. It
is, however, not possible to trace them at this stage across to the
opposite side of the brain. There is as yet no chiasma. The two
indentations *(CH., CH.)* are caused by the choroidal fissure.

From the end of the cleft nearest the brain the fibres of the optic
nerve are seen issuing *(OP. N.)*, and alongside the nerve, but quite
separate from it, is the optic stalk *(OP. S.)*, the lumen of which is
continuous with the cavity of the primary optic vesicle at *C. OP. V.*
In the next few sections fibres of the optic nerve can be traced into
connection with those arising from cells within the retina, such as are
seen at *N. F.*

Fig. 2 is a camera drawing of the optic stalk and nerve of the same
tadpole, but of the left instead of the right side. In this figure the
optic nerve and stalk, cut across at right angles to their longitudinal
axes, are seen lying between the band of dense tissue *(TR. CH.)* and
the brain *(BR.)*.

The evidence here, again, is to prove the optic nerve to be entirely
independent of the optic stalk, except that the two structures lie
closely opposed to one another.

A section transverse to the longitudinal axis of the nerve-stalk,
taken at any point between the eye and brain at this stage, would
give a similar figure.

The cells of the wall of the stalk on the side on which the nerve
lies are as large and in every way similar to those on the opposite side,
whence it is impossible to maintain that the nerve-fibres are developed
by the differentiation of the cells of the stalk *in situ* into nerve-fibres.

Fig. 3 shows the connection of the optic nerve and stalk with the brain.

On reaching the brain the stalk and nerve separate, the cavity of
the optic stalk becoming continuous with the optic recess in the floor
of the third ventricle, while the fibres of the optic nerve can be traced
as far as the middle line across the great ventral commissure *(M. COM.)*,
but at this stage can be traced no further. In later stages (20 mm.)
the fibres may be easily traced to the opposite side of the brain, and
later (30—40 mm.) up into the optic lobe of the opposite side.

At the stage I have been describing the optic chiasma cannot be
said to exist ; or if some of the fibres have indeed crossed, they form
such small bundles that they cannot be recognised.

Figs. 5, 6, 7 represent much earlier stages (tadpole, 7 mm.), by which time the nerve-fibres have just made their appearance, and figs. 8, 9, 10, 11 represent considerably older stages (tadpole, 23 mm.).

Figs. 5, 6, 7 are placed with their dorsal surfaces towards the top of the page, and are figures of sections taken close to the eye.

Figs. 1, 2, 3 are placed with their anterior surfaces towards the top of the page, and are figures of sections taken close to the eye, midway between the eye and brain, and close to the brain, respectively.

Figs. 8, 9 are placed with their dorsal surfaces towards the top of the page, and are figures of sections taken close to the brain.

From a study of these figures it will be seen that the optic nerve lies at its retinal end ventral to the stalk ; but as it nears its cerebral end it lies along the posterior border of the stalk, and even in later stages (fig. 8) lies almost dorsal to the stalk.

The history of the relation of the stalk to the nerve in later stages is as follows :—In tadpoles of 11—12 mm. the trabecula cranii cartilage growing up under the optic stalk and nerve causes the lumen of the stalk to become obliterated at that point, from which point the obliteration of the whole lumen gradually proceeds.   The last part to remain open is the part nearest to the brain, which in tadpoles of 23 mm. is still open (fig. 8).

As the nerve-fibres increase in number they seem to tend to grow in between the cells of the walls of the stalk (fig. 9), and eventually (fig. 10) the walls of the stalk become completely broken up, and the cells remain separated from one another, and lie among the fibres of the optic nerve as in fig. 10.   Fig. 11 is a longitudinal vertical section of the optic nerve of a 23 mm. tadpole.   Very possibly this breaking of the optic stalk is caused not so much, if at all, by the nerve-fibres growing in amongst the cells, but, primarily, by reason of increase in distance between the eye and brain, the stalk becomes stretched and broken up.

It might be argued that the nerve-fibres grow out of processes from the cells of the stalk, but of this there is no trace at any time.   The cells of the stalk never show any processes, such as are easily and distinctly seen in the neuroblasts of the rest of the nervous system, and excepting a considerable diminution in size (which is common to almost all cells of all parts of the animal), the cells of the optic stalk of a 10 mm. tadpole do not materially differ in shape or character from those of the tadpoles in which nerve-fibres have not yet appeared.

I cannot see that there can be the slightest doubt that the nerve-fibres of the optic nerve are developed from some part other than the optic stalk. Hence we are bound, I think, to conclude that the fibres must be outgrowths from cells in either the brain or in the retina.

Origin of the fibres of the Optic Nerve.

I believe that with the exception of Froriep's (7) researches mentioned at the beginning of this paper, and Keibel's (13) short statement, and of His's (12) expressions of opinion, there is no actual evidence published of the growth of the fibres along the optic stalk, either towards the brain or towards the eye.

His's (12) evidence, though perhaps not complete, is very strong, for he finds in human embryos of five weeks (13 mm.) the first trace of nerve-fibres in the retina as follows :

" Ich selber finde die ersten Nervenfasern der Retina bei menschlichen Embryonen von etwa 5 Wochen (13 mm.). Zu der Zeit zeigt die Retina an ihrer Innerseite ein von den Müllerschen Fasern durchzogenes weites Raumsystem. Die Fasern bilden eine erst dünne, scharf auslaufende Schicht, und sie treten unter spitzen Winkeln aus der austossenden Zellenlage hervor. Hier liegen kleine retortförmig gestaltete Neuroblasten, die mit ihren umgebogenen Spitzen in die Fasern sich fortsetzen. Die zuerst gebildeten Opticusfasern entstammen somit den Zellen der Retina und wachsen centralwärts. In die inneren Körnschicht dagegen finden sich zahlreiche Neuroblasten welche ihre Spitzen nach auswärts kehren."

All three authors, however, agree that the growth of the nerve-fibres is from the eye to the brain.

In support of their views I may add the result of my own observations on the frog.

Before explaining my figures on this point I must mention the beautiful work of Cajal (3).

The author, who made use of the Golgi and Weigert-Pal method, describes the result of his observations on the optic lobes of adult birds and of advanced embryos, amongst which were chicks of the tenth day of incubation and upwards.

Cajal finds that the great majority of the fibres of the optic nerve end in the optic lobes, in many fine-branched twig-like endings, but in no ▯▯▯ ▯▯ these ending anastomose or become directly connected with nerve-cells.

In an earlier paper (4) he described the termination of the nerve-fibres in the retina as of two kinds. I quote from the first-named paper (3), in which he on p. 338 gives a short account of his previous work :

" Dans cette membrane les fibres du nerf optique se terminent de deux façons ; par les cellules, c'est à dire, se continuant avec les cylindres-axes des éléments de la couche ganglionaire, et par des ramilles variqueuses et indépendentes situées à la rencontre des couches réticulaire interne et des grains internes. De ces deux espèces de terminaisons, seules les dernières peuvent ainsi se qualifier, car les premières ne sont point de véritables terminaisons, mais plutôt des origines de fibres dont la fin arborisée doit se trouver dans le lobe optique. Si la doctrine de l'indépendance des éléments nerveux est certaine, le lobe optique devra nous montrer, de même que la rétine, des cellules d'origine, et des arborisations terminales."

Cajal (3), in what he calls the tenth layer (couche 10, equivalent to ▯▯▯ ninth granular layer of Stieda [24] and fifth zone or zone fusiform ▯▯▯▯ ▯▯ Bellonci [2]), finds cells of an ovoid or spherical shape with a protoplasmic expansion running inwards and outwards. The inward or inferior one seems to end quickly, while the outer or superior runs outwards as far as the layer of fibres of the optic nerve. From the latter process arises an axis-cylinder which enters the layer of optic nerve-fibres—couche 1 of Cajal. With reference to these cells, and fibres arising from them, Cajal says on p. 248,—

Comme nous venons de le voir ici déjà nous rencontrons les origines cellulaires de quelques fibres optiques lesquelles, d'après nos recherches sur le rétine des oiseaux pourraient bien être celles qui finissent dans la couche des grains internes par des arborisations libres, courtes et fortement variqueuses. Nous ignorons si, parmi les fibres du nerf optique, il en existent d'autres d'origine centrale."

Hence it seems almost certain from the study of the adult condition, at any rate in birds, that the optic nerve is composed of ▯▯▯ kinds : (1) those which have ▯▯▯▯▯ ▯▯▯▯ ▯▯▯▯ in the ▯▯▯▯▯ ▯▯▯ ▯▯▯▯

grown centralwards; (2) those which have arisen in the brain and grown outwards.

From a study of the development of the eye and optic nerve of *Rana temporaria*, I am convinced that at any rate a very large proportion of the nerve-fibres of the optic nerve arise as outgrowths of cells in that portion of the optic cup from which the retina will be formed, which processes grow centralwards at first along the ventral, then along the posterior border of the optic stalk, and entering the brain immediately posterior to the optic recess cross along the ventral surface of the middle commissure to the opposite side, where they turn dorsalwards and slightly backwards to the roof of the mid-brain.

In tadpoles of about 7 mm. in length, or a little earlier, the first trace of the optic nerve may be seen.

Many of the cells of the retinal portion of the optic cup may be seen to be pear-shaped, with their end drawn out into processes which are directed towards the centre of the optic cup.

It is not possible to follow the individual fibres, but all are directed towards the ventral rim of the optic cup, over which, at any rate, a large number pass and run along the ventral border of the optic stalk, some for a greater, others for a lesser distance.

Fig. 4 is from a sagittal section of a tadpole (of 7½ mm. in length), and therefore a section transverse to the longitudinal axis of the optic stalk, taken at the point where the fibres are crossing the ventral lip of the optic cup.

It will be seen that the fibres pass through a cleft, or rather, I should say, the walls of the optic cup have grown up round the bundle of nerve-fibres since they passed over the rim, for there is no such apparent cleft before the fibres are developed. Fig. 5 is a section of the optic stalk from the same series as Fig. 4, but taken between the eye and brain, close to the former. Here on the ventral border is seen the bundle of fibres, and the wall of the optic stalk is slightly bulged in along the line of the fibres. The bundle here is rather smaller than in Fig. 5.

In Fig. 6, which is also from the same series of sections, but taken nearer the brain, the bulging in of the optic stalk is not so marked, and the bundle of fibres considerably diminished in size. Between this section and the brain the fibres become less and less distinct, and in sections of the optic stalk close to the brain no trace of nerve-fibres can be seen (Fig. 7). Such is the case in tadpoles of 7—8 mm. in length. In tadpoles of 8—9 mm. the nerve-fibres can be seen along the whole of the stalk.

In later stages fibres may be traced to the upper regions of the mid brain, though of the actual terminations I have nothing to say. Possibly the appearance of concentric bands of white matter of a molecular appearance in the roof of the optic lobes is concomitant with the branching of the terminations of the fibres that have reached that part of the brain ; but, as far as my own researches go, that is merely conjecture. These molecular layers are first visible in tad poles of 17—20 mm. in length. I consider, however, that the fibres have reached the upper part of the optic lobes in younger tadpoles— those of 12—15 mm.

The optic fibres begin to grow as soon as those from any other part of the nervous system, and grow very rapidly. As the eye itself increases in size the bundle of fibres lying along the optic stalk increases also.

The fibres whose addition causes the increase in size of the bundle arise as outgrowths from neuroblasts (Fig. 13, *N*., Plate IX.) near the rim of the optic cup, for along this rim a constant proliferation of cells takes place as long as the eye increases in size.

Fig. 13 shows the main features in the structure of the retina of a tadpole of 13 mm., and is as far as possible a camera drawing.

Towards the centre of the cup the development of the retina is furthest advanced.

The two walls of the optic vesicle are closely approximated, and the originally inner wall is reduced to a single layer of pigmented cells (*P.*).

These pigmented cells are directly continuous with the walls of the optic stalk, and thereby with the epithelial lining of the neural tube. These cells represent the epidermic layer of the epiblast, so that from this part of the optic cup it will be seen that the nervous layer is entirely absent.

In the originally outer wall of the optic vesicle there are many cells derived from the nervous layer, and the epidermic cells are greatly elongated and to a certain extent branched, and form the supporting elements only of the retina.

The relation of these two parts is shown diagrammatically in fig. 12, which represents a rather earlier stage (tadpole 9 mm.).

The further discussion as to the fate of the nervous and epidermic layers of epiblast I defer to a paper on the development of the central nervous system of the frog, on which I am at present engaged.

I think, however, that there is no doubt that the spongioblastic elements are all derived from the epidermic layer, and the neuroblastic from the nervous layer of the primitive epiblast.

To return to the section. Next to the pigment layer are seen the developing rods and cones. The former are more developed than the latter, and already their processes, which project into the cavity of the optic vesicle, show the division into inner and outer limbs.

Connected with the outer molecular layer are certain neuroblasts of the inner nuclear layer (X).

Among the cells of the outer nuclear layer are some with long, broad prolongations (FR) the radial or Müllerian fibres of the adult.

These fibres are not easily traced to the "outer limiting membrane" in the centre of cup, but may be more easily seen towards the rims, where the rods and cones are not so crowded (FR'). Compare these with the cells marked SP in fig. 12.

Certain neuroblasts (XI) of the inner nuclear layer seem to send processes into the inner molecular layer, but my observations do not enable me to say whether they break up into fibrils, or whether they pass through and on to the brain.

Within the inner molecular layer comes a double row of cells, the ganglionic layer. In many cases these cells are produced into short deeply staining processes towards the inner molecular layer, and break up at once into fine fibrils (G.).

The growth of these processes seems to be the prime cause of the appearance of the inner molecular layer.

These processes just described appear to be a later development than the process from the opposite pole, which forms a fibre of the optic nerve.

These are seen at N, near the rim of the cup, as thick black prolongations, while the other pole is still round, and shows as yet no signs of the processes described above, which form, at any rate in part, the inner molecular layer.

So also if the other elements of the eye be traced through the section, it can be noticed that as they near the rim they are less and less differentiated until they all merge into a mass of rapidly dividing cells, each one very like to his neighbour.

The Choroidal Fissure.

At one point of the rim, however, there is no proliferation of cells, and therefore at this point the wall of the optic cup does not grow. This point is that over which the fibres from the retinal neuroblasts pass on their way towards the brain (fig. 12, *CH*).

It is quite inaccurate to talk of the fibres of the optic nerve becoming connected with the elements of the inner or retinal wall of the cup after piercing the outer wall of pigment epithelium (Foster, 6), as the development shows that the fibres never really pierce either wall, but, from the moment of their first formation, they are on the outside of both. It is only by the subsequent growth of the rim of the optic cup that the bundle of nerve-fibres becomes surrounded by the walls of the cup, and so apparently pierces it. It does in reality pass over the edge of the cup, just as much as do the fibres in such an eye as that of Pecten.

It has been usual to regard the choroidal fissure as essentially an embryonic feature, present chiefly for the purpose of admitting mesoblastic tissues into the optic cup for the formation of and nourishment of the vitreous body, and to be due to the manner of invagination of the optic vesicle.

Some authors have recognised a further meaning in that the optic nerve is thereby brought into connection with the retina (v. Hertwig, 9, p. 404).

I have never, however, seen it suggested that the choroidal fissure represents a stage in the evolution of the eye, as seems to me more than probable, and that it was due entirely to the eye having a deep-seated cerebral origin, and having only subsequently grown towards the surface.

Whatever may have been the first origin of the eyes of Vertebrates, whether they arose, as has been suggested by Balfour (1), as patches of the epidermis sensitive to light, before the sinking down and folding up of the central nervous system, or whether, as Lankester (15, 16) suggested, they are derived from such an eye as that found within the cerebral vesicles of certain Ascidians, it is clear that they were of myelonic origin, and much more deeply placed than at present in adult Vertebrates.

In either case the light must have fallen directly upon the sensitive cell; that is to say, the light reached the eye from the opposite direction to what it does in Vertebrate eyes of the present day.

That is to say, the light-perceiving portion of the sensory cell would be directed towards the cavity of the brain, and the transmitting portion or nerve-fibre towards the exterior, as indeed is the case in the larval Ascidian eye.

When we realise that the nerve-fibres of the present Vertebrate eye really pass over the edge of the cup, and do not—morphologically speaking—pierce it, we are able to imagine the probable steps in the evolution of the eye far more easily than otherwise.

It can hardly be supposed that such eyes perceived any image; it was merely a case of distinguishing light from dark. On the closure of the neural tube in the one case, or on the commencement of opacity in the other, and more light therefore reaching the light-sensitive cells from the opposite direction to that heretofore, any variation (1) which brought the sensory patch nearer the skin (origin of optic vesicle), (2) which brought the skin nearer the sensory patch, i.e., depression in the skin (origin of lens), would tend to be preserved.

As yet the eye would not be a cup; it would only become so in connection with the formation of the lens.

Round the depression in the skin the light sensitive area might expand, and by the growth of its edges round the depression in the skin would form a cup.

While these changes were in progress the nerve-fibres having now to pass over one part of the edge of the area to reach their cerebral destination, would prevent the growth of the edge at that point, and consequently a gap would be left.

As soon as a lens was formed and an image thrown upon the retina, a gap would be disadvantageous to the perception of the image as well as to the retention of the vitreous body, which no doubt existed as early as the lens; but until such a time I do not see why a choroid fissure should not be a permanent feature; and indeed it seems to me that a consideration of the manner in which the Vertebrate eye was evolved almost necessitates the occurrence of a gap at a certain stage.

Development of Optic Nerve in Chick.

I have followed also the development of the optic nerve in the chick, and find that the mode of development is essentially similar to that described above for the frog, *Rana temporaria*.

In chicks of four days thick nerve-fibres may be found in the retina, radiating towards and into the just beginning choroidal fissure, but can be traced no further. In five-day chicks the fibres are thinner, and can be easily traced into the choroidal fissure, but along the optic stalk near to the brain there is no trace of nerve fibres.

In six days fibres can be traced all the way to the brain.

SUMMARY.

1. The optic stalk takes no part in the formation of the nervous parts of the organ of sight.

2. The optic stalk becomes broken down and the cells composing it separated from one another, partly by the mechanical stretching due to the growth of the optic nerve, partly by the growth in between the several cells of the nerve-fibres.

3. The optic nerve is developed independently of the optic stalk, the nerve-fibres lying along the posterior border of the stalk, and at first entirely outside it ; but on the breaking down of the stalk some of the nerve-fibres grow in between the cells.

4. The great majority of fibres forming the optic nerve arise as outgrowths from nerve-cells in the retina, and grow towards and into the brain.

5. According to Cajal's researches certain fibres also exist which would seem to grow from the central nervous system to the retina, but these I have not been able to find.

6. The nerve-fibres pass over the ventral edge of the optic cup, and thereby cause the formation of the choroidal fissure.

7. The choroidal fissure of the embyro represents a condition in the evolution of the eye which was persistent in the adult prior to the formation of a lens.

M

It has only secondarily been made use of as a means of ingress for the mesoblastic tissues.

ALPHABETICAL LIST OF LITERATURE REFERRED TO.

1. BALFOUR.—' A Treatise on Comparative Embryology,' 2nd edition, 1885.

2. BELLONCI.—" Ueber die centrale Endigung des Nervus Opticus bei Vertebraten," ' Zeitschrift für wiss. Zool.,' Bd. xlvii, 1888.

3. CAJAL, RAMON Y.—"Sur la fine structure du lobe optique des oiseaux, et sur l'origine réelle des nerfs optiques," ' Monthly International Journal of Anatomy and Physiology,' vol. viii, parts 9 and 10, 1891.

4. CAJAL, RAMON Y.—"Sur la morphologie et les connexions des éléments de la rétine des oiseaux," ' Anatomischer Anzeiger,' No. 4, 1889.

5. DOHRN.—' Ursprung der Wirbelthiere," 1875.

6. FOSTER.—' A Text-book of Physiology,' 5th edition, M. Foster, 1888—1891.

7. FRORIEP, A.—' Anatomischer Anzeiger,' vi, 1891.

8. HADDON, A. C.—' An Introduction to the Study of Embryology,' 1887.

9. HERTWIG, O.—' Lehrbuch der Entwickelungsgeschichte des Menschen und der Wirbelthiere,' 1890.

10. HIS.—' Untersuchungen über die erste Anlage des Wirbelthierleibes,' Leipzig, 1868.

11. HIS.—" Die Neuroblasten und deren Enstehung in embryonalen Mark," ' Archiv Anat. u. Phys.,' 1889.

12. HIS.—" Histogenese und Zusammenhang der Nerven-elemente," ' Archiv für Anat. und Phys.,' 1890, Supplement.

13. KEIBEL.—' Deutsche medicinische Wochenschrift,' 7th Feb., 1889.

14. KÖLLIKER.—' Entwickelungsgeschichte der Menschen und der hoheren Thiere,' Leipzig, 1879.

14a. KÖLLIKER.—' Zur Entwickelung des Auges und Geruchsorgans menschliche Embryonen. Zum Jubiläum der Univ. Zurich,' Würzburg, 1883.

15. LANKESTER, E. R.—' Degeneration,' 1880.

16. LANKESTER, E. R.—' The Advancement of Science,' 1890.

17. LIEBERKÜHN.—" Beiträge zur Anatomie des embryonalen Auges," ' Archiv f. Anat. und Phys.,' 1879.

18. LIEBERKÜHN.—' Ueber der Auge des Wirbelthier Embryo,' Cassel, 1872, v. Hertwig, 446.

19. MARSHALL.—' The Frog,' A. Milnes Marshall, 1891.

20. MÜLLER, W.—' Ueber die Stammesentwickelung des Sehorgans der Wirbelthiere,' Leipzig, 1874.

21. ORR.—"Contribution to the Embryology of the Lizard," 'Journal of Morphology,' vol. i, No. 2, 1887.

22. ORR.—" Note on the Development of Amphibians," 'Quart. Journ. Micr. Sci.,' vol. xxix, 1887.

23. SCHAFER.—" Embryology," Part 1 of vol. i of 'Quain's Elements of Anatomy,' 10th edition, 1890.

24. STIEDA.—"Studien über das centrale Nervensystem der Vögel und Saugethiere," 'Zeits. f. wiss. Zool.,' Bd. xxx, 1868.

25. SPENCER, W. B.—"On the Presence and Structure of the Pineal Eye in Lacertilia," 'Quart. Journ. Micr. Sci.,' 1886.

EXPLANATION OF PLATES VIII. & IX.

Illustrating Mr. Richard Assheton's paper "On the Development of the Optic Nerve of Vertebrates and the Choroidal Fissure of Embryonic Life."

Alphabetical List of Reference Letters for all the Figures.

BR. Brain. *B. V.* Blood-vessel. *C.* Cone. *C. OP. V.* Cavity of the optic vesicle. *CH.* Choroid fissure. *CH'.* Choroid fissure, outer end. *D.* Growing rim of optic cup. *FR.* Radial fibres. *G.* Cells of the ganglionic layer of the retina. *L.* Lens. *L. C.* Cavity of lens. *M. COM.* Middle commissure. *MES.* Mesoblastic sheath round optic nerve and stalk. *MO.* Outer molecular layer. *N.* Neuroblast. *NI.* Inner nuclear layer. *O.P. N.* Optic nerve. *OP. RC.* Optic recess. *OP. S.* Optic stalk. *OP. S'.* Cells of optic stalk separated. *P.* Pigment layer of retina, i.e., posterior wall of optic vesicle. *R.* Rods. *R. CH.* Edge of choroid fissure. *RS.* Rim of optic cup. *TR. CR.* Trabecula cranii. *X.* Neuroblast of inner nuclear layer.

FIG. 1.—Horizontal section of the right eye of a 10 mm. tadpole.

FIG. 2.—From a horizontal section of a 10 mm. tadpole. In this the optic stalk *OP. S.*, and optic nerve *OP. N.*, are cut across transversely to their length.

FIG. 3.—From a horizontal section of a 10 mm. tadpole. The optic nerve and optic stalk are seen separating at their junction with the brain.

FIG. 4.—From a sagittal section of a 7 mm. tadpole. The figure shows the ventral edge of the posterior part of the eyeball; the optic nerve is seen passing through the choroidal fissure.

FIG. 5.—From the same series as Fig. 4, taken nearer to the brain. 7 mm. tadpole.

FIG. 6.—From the same series as Figs. 4 and 5, taken about midway between eye and brain, nearer to brain than Fig. 5. 7 mm. tadpole.

Fig. 7.—From the same series as Figs. 4, 5, and 6, taken still nearer to the brain.

Fig. 8.—From a sagittal section of a 23 mm. tadpole, taken near to the brain. Optic nerve and optic stalk cut transversely.

Fig. 9.—From a sagittal section of a 23 mm. tadpole, as Fig. 8.

Fig. 10.—Same as Figs. 8 and 9, but taken nearer to the eye.

Fig. 11.—From a transverse section of a 23 mm. tadpole. The optic nerve is cut "sagittally;" the cells of the broken-up optic stalk are seen scattered over it.

Fig. 12.—Semi-diagrammatic figure of a solid section of an eye of an 8 mm. tadpole. The optic nerve-fibres are seen on the cut surface to be processes of neuroblasts (blue). They pass over the ventral edge of the optic cup (the choroidal fissure). The edge of the choroidal fissure is seen at *E. CH.*, the mosaic-like pattern being the pigment cells of the hinder wall of the optic vesicle. The optic stalk is seen to be hollow, and quite separate from the optic nerve. The cells of the walls of the optic stalk pass into the pigmented cells of the outer wall, and the spongioblasts of the inner wall of the optic cup, *SP.* The blood-vessel which enters the optic cup is seen cut off at *B. V.*

Fig. 13.—A section of the retina of a 13 mm. tadpole. A description of this figure is given in the text. Prep. Perenyi's fluid ; aniline blue-black.

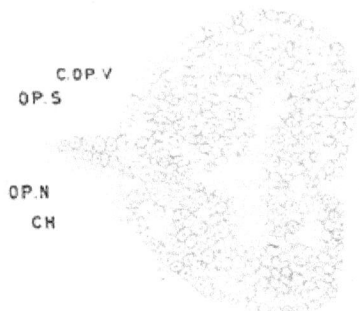

Fig. 1.

C.OP.V
OP.S

OP.N
CH

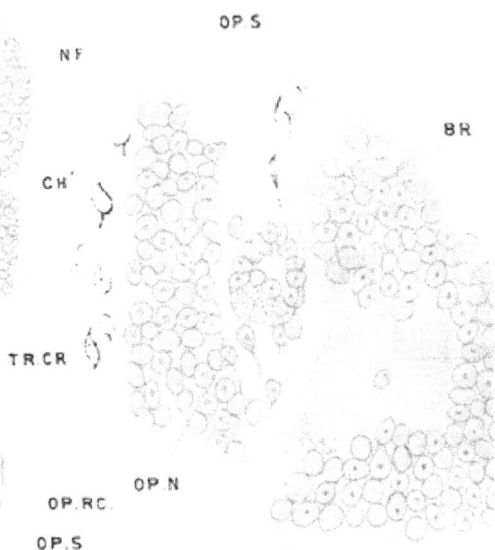

Fig. 2.

OP.S

N F

BR

CH'

TR.CR

OP.N

OP.RC.

OP.S

Fig. 3.

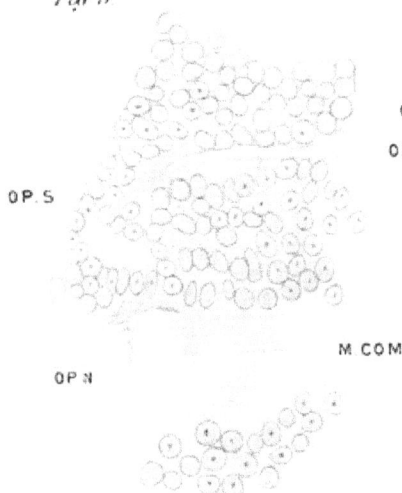

OP.S

OP.N

M.COM

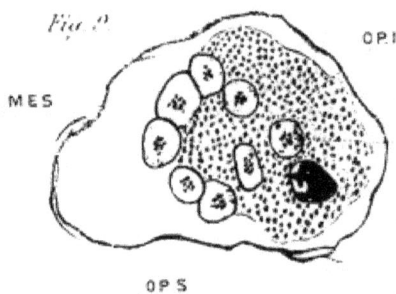

Fig. 4.

MES

OP.I

OP.S

Fig. 5.

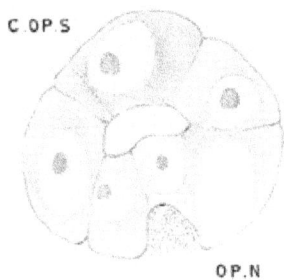

C.OP.S

OP.N

Fig. 6.

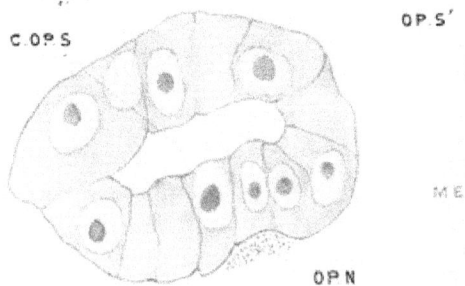

C.OP.S

OP.S'

MES

OP.N

Fig. 7

OP S

M.COM

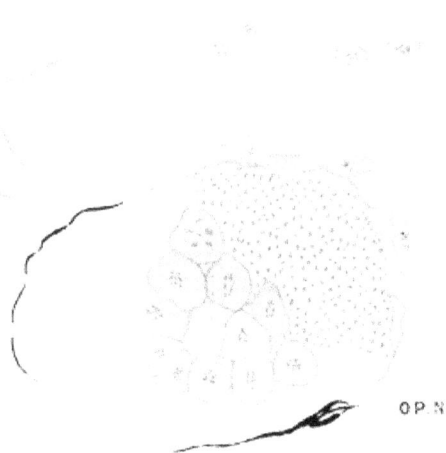

Fig. 8   BR

OP.N

OP N   MES   OP S

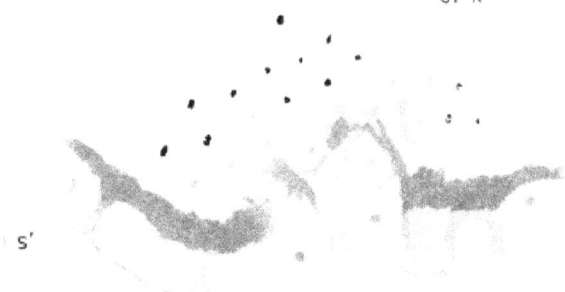

Fig. 9

S'

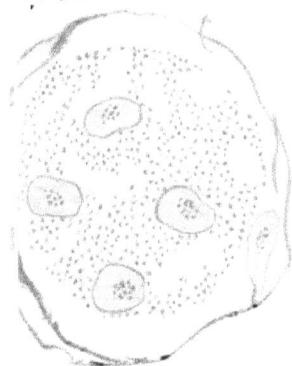

Fig. 10

CH

OP S'

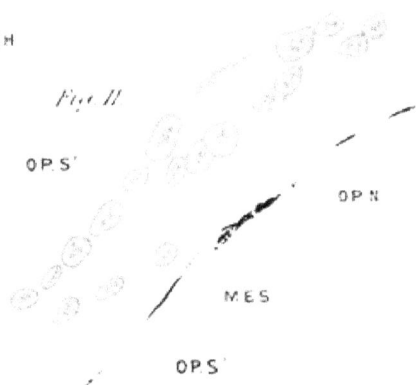

Fig. 11

OP S'

OP N

MES

OP S'

Plate iX

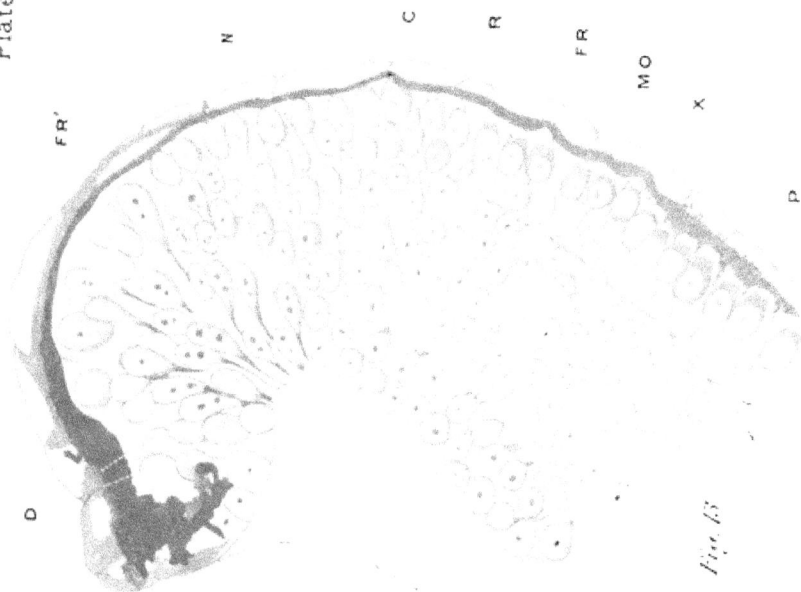

Fig. 13.

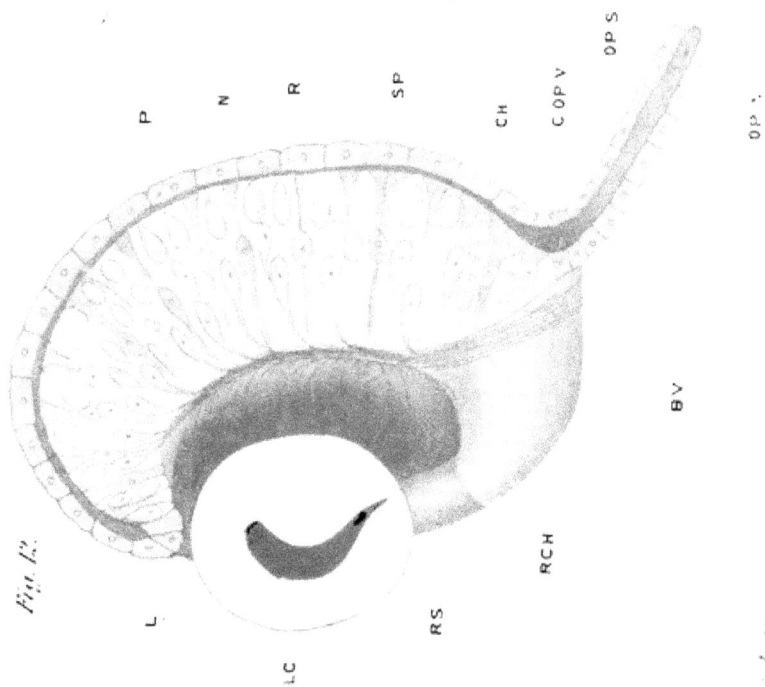

Fig. 12.

Reprinted from the QUARTERLY JOURNAL OF MICROSCOPICAL SCIENCE, *April, 1893.*

CONTRIBUTIONS TO A KNOWLEDGE OF BRITISH MARINE TURBELLARIA.

By F. W. GAMBLE, B.Sc., *Berkeley Fellow of Owens College, Manchester.*

With Plates X, XI, & XII.

CONTENTS.

I. INTRODUCTORY.

1. *Nature and Scope of the Paper.*

It is usual to begin a faunistic paper on Turbellaria with the words of Oscar Schmidt, "Man braucht, wie es scheint, wo man will nur zuzugreifen und ist der Ausbeute sicher ;" and although written forty years ago, they are still applicable to this group. In the northern European seas (and it was to these that Schmidt referred) many species may be found by careful examination of almost any portion of the littoral and laminarian zones. Yet, owing probably to their small size, inconspicuous appearance, perishable nature, and obscure anatomy, few naturalists in this country have devoted much attention to them. When once a keen interest, however, is aroused, the number of new and morphologically important forms that may be found in a limited area is beyond anticipation. This has been well expressed by Jensen in the preface to his work on the Turbellaria of the west coast of Norway : " The new forms described here are only a very small part

of those that occur on our coasts. A rich harvest remains behind.
Indeed, one has no need to go far to find numbers of new species.
Among seaweed they are frequently brought up at every haul.
At greater depths, again, other new species are found ; but as regards
the occurence of the smaller and more numerous forms, our fauna is
still quite unknown." *

This quotation applies very appositely to the present case. As yet
only a small portion of our coast has been explored. Extended
observations are urgently needed. Still, from my own list, together
with the records of other naturalists who have from time to time
noted the occurrence of British forms, it appears that our fauna
already includes fourteen Polyclads, two Triclads, and fifty-five
Rhabdocœles—a total of over seventy species. It is to the description
of these that I address myself. In doing so I shall try to indicate the
distinctive points, structural and bionomical, that separate the various
forms, reserving a more detailed account of the anatomy of new
or partially known species to a future paper.

The specific descriptions (which largely confirm those of previous
observers) are taken, except where otherwise stated, from my own
observations.

One section of the group—the parasitic Turbellaria—is omitted.
These forms undoubtedly occur on our coasts, but I have had no
opportunity for their investigation.

2. Historical Account.

The history of British marine Turbellaria may be said to begin with
the publication of Dalyell's octavo volume, "Observations on some
Interesting Phænomena in Animal Physiology exhibited by Several
Species of Planariæ" (1811 and 1814). Among the eight species
there described, is a marine one, Planaria flexilis (now known as
Leptoplana tremellaris), from the Firth of Forth. This animal, how-

* Preface : " De her beskrevne nye Former ere da kun en meget liden
Brökdel af de ved vore Kyster forekommende. En rig Höst staar tilbage.
Man behöver sandelig ikke at gaa langt for at finde en Mængde nye Arter. I
Tangen i Fjæren kan man jevnlig drage saadanne op hvert Kast . . . . . men
Faunan her, navnlig for de mindre og talrigste Arters Vorkommende, er endnu
helt unbekjendt " (49). The numbers in brackets refer to the list of literature
at the end of this paper.

ever, was not new. It had been discovered and carefully described by
O. F. Müller, in his 'Vermium terrestrium et fluviatilium Historia,'
nearly forty years before. While, therefore, Dalyell made no new
discovery in *Planaria flexilis*, his study of its habits considerably
increased our existing knowledge.

The merit of his work lies in the careful and patient observations,
the accumulation of many years, which it records. The muddy
haunts of the *flexilis*, its active behaviour when in search of food, its
inordinate appetite after a period of starvation (illustrated by a case
in which the bodies of three individuals burst and subsequently
putrefied, owing to the quantity of absorbed food), the increase in
bulk and changes of colour due to the contained nutriment, the mode
of propagation by eggs laid in batches like those of molluscs, and the
power of repairing serious injury,—all these points are graphically
described and extended to seven fresh-water forms. In fact, this book
contains even at the present time the best account that we possess of
the bionomics of these animals, and well earned for Dalyell the
dedication by v. Graff of his 'Monographie der Turbellarien.'

From 1814 to 1852 the work done on this group was confined, in
this country, to the description of single forms, or to the records of
their occurrence on our coasts. Montagu (7) in 1815 discovered
Planaria vittata (*Prostheceraeus vittatus*) on the south coast of Devonshire. In 1821 Fleming (8) found *Planaria atomata, tremellaris*, and
vittata during a voyage round Scotland. Our knowledge of this part
of the British fauna was further increased by a series of papers by Dr.
George Johnson, and Wm. Thompson, of Dublin. The former, under
the title 'Illustrations of British Zoology' (11 and 12), described
Planaria cornuta (*Eurylepta cornuta*) and *Planaria subauriculata*
(*Stylochoplana subauriculata*) from the shores of Berwick Bay. These
accounts suffice to enable us to identify the species referred to, but
give little idea of their internal anatomy. The relations of the known
forms to one another were as yet quite obscure. Thompson recorded
forms from various Irish localities. Forbes and Goodsir found
Polyclads in the Orkneys and Shetlands in 1839 (13), and in his
later dredging reports Forbes frequently enters "*Planaria sp.?*"
accompanied by a lament that so little is yet known of these forms.

If we turn to work done on the Continent during this period

(1830—1850) we find that v. Baer and Dugès had independently discovered the internal anatomy and especially the generative organs of the fresh-water Planariæ, and had proposed a rough scheme of classification. The corresponding discovery of the anatomy of Polyclads was made by v. Mertens.* Ehrenberg (10) had found many new forms, some of which constituted his subdivision " Rhabdocœla " Oersted's most important paper (16) appeared in 1844,† containing a system which included all known species, and from which later attempts originate. Our knowledge of the histology and anatomy of the Rhabdocœles (the small size of which usually prevented an adequate investigation being made, on account of the necessity for high magnifying powers) was immensely increased by Max Schultze's ' Beiträge zur Naturgeschichte d. Turbellarien,' 1851. The accuracy of the descriptions and beauty of the copperplates are well known.

The following year (1852) Dalyell published the ' Powers of the Creator,' the second volume of which contains references to a considerable number of Turbellaria. Among these, *Monotus lineatus* and *Convoluta paradoxa* are interesting as being the first records of British Rhabdocœlida. The habits and reproduction are well described, but the anatomy is very far behind the knowledge of the time. For ten years (1852—1862) little work on these forms was done in this country, while Oscar Schmidt, Schultze, and Leuckart on the Continent were extending a monographic and systematic knowledge of the group. In 1859, however, Claparède spent August and September in the Hebrides, chiefly at Skye. In a most interesting paper (35) he describes *Convoluta paradoxa* (in which he determined successive hermaphroditism), *Mesostomum marmoratum, Prostomum caledonicum, Vortex quadrioculata, Euterostomum fingalianum,* and the Polyclads *Centrostomum Mertensii* and *Eurylepta aurita.* Similar researches (36) on the coast of Normandy showed what a varied Turbellarian fauna existed there. No one, however, was found in this country to advance our knowledge of the group on similar lines.

In 1865 the ' Catalogue of Non-parasitical Worms in the British

---

* " Untersuchungen ü. d. Bau verschiedener in d. See lebender Planarien," ' Mém. de l'Acad. Imp. d. Sciences d. St. Pétersbourg,' sér. 6ᵐᵉ, t. ii, 1833.

† Reprinted with figures and additions from ' Kröyers Naturhist. Tidsskrift,' 1843.

Museum' appeared. The marine Turbellaria are taken from the works of Johnston, Thompson, and Dalyell, together with a few new records.

A year later Lankester (39) issued a list of the fauna of Firman Bay, Guernsey, containing *Convoluta paradoxa*, *Leptoplana auricularis*, *L. flexilis*, and *Eurylepta cornuta*. In 1875 McIntosh (45) published his 'Marine Invertebrates and Fishes of St. Andrews,' in which several Turbellaria are mentioned. The occurrence of *Prostoma lineare* Oe. (*Gyrator hermaphroditus*, Ehrg.), in the sea, and a short description of *Mesostoma bifidum*, n sp. (*Pseudorhynchus bifidus*), are specially note-worthy. v. Graff paid a visit to Millport, the result of which are incorporated in his great monograph ([53,] 1882, p. 437). Twenty-four marine species of this group were found and fully described. In the summers of 1884 and 1885, Koehler explored the Channel Islands. A list of the forms obtained may be found in the 'Annals and Mag. of Nat. Hist.,' fifth series, vol. xviii, p. 362. The most important additions to the Turbellarian fauna are *Oligocladus sanguinolentus* and *Proceros argus*.

3. *Nomenclature.*

I wish in this section to discuss certain difficulties connected with the terminology of the complicated reproductive organs of the Turbellaria.

The stages exhibited by different members of this group, by which a simple organ, producing ova capable of manufacturing the necessary food-yolk, becomes differentiated into two parts, one furnishing ova, the other yolk, and the final separation of these two parts into two distinct organs, have been pointed out by Gegenbaur, Balfour, and others. Thus in the *Acœla* the organ is quite simple, the ova elaborating their own food-yolk. To such an organ I shall apply the term *ovary*. Certain *Rhabdocœla—e. g.* Prorhynchus—exhibit the first stage in complexity. The cells are still equivalent, but are not all equally capable of becoming ova; those that are not, form yolk-cells destined for the nutrition of the ovarian part of the organ. The secretion of yolk-granules by the yolk-cells surrounding the fertilised ova, which in this form does not take place until the ova have undergone segmentation, in the *Cylindrostomina* and others occur before the yolk is transferred to the ova. To such an

organ, performing a double function, the Germans apply the term
"Keimdotterstock," and as an equivalent I shall use *"germ-yolk-gland,"*
although, strictly speaking, the word "gland" should apply to the
vitelline portion only.

In the great majority of Turbellaria the two parts become separate
organs with distinct functions. For these I use the old terms
germarium for the ovarian organ, and *vitellarium* or *yolk-gland* for the
nutritive one. The term *vagina* I apply to that part of the female
genital duct which forms a sheath for the penis during copulation.
For the storage and nutrition of the spermatozoa various accessory
organs are developed. For a single organ, serving to retain the sperm
until fertilisation is accomplished, the term *spermotheca* is a convenient
equivalent for "bursa seminalis," used by v. Graff. In many cases
(*e.g.*, Vorticidæ) two organs are present, one of which receives the male
products of another individual, and then passes it on to the second,
from which fertilisation takes place. I retain the term *bursa
copulatrix* for the former muscular structure, and *receptaculum seminis*
for the latter. While the ova are being duly fertilised, provided with
food-yolk, and surrounded by an egg-capsule, they are usually retained
within the body of the parent during, and usually also a short time
after, these changes. Consequently a certain amount of development
is passed through. To the region in which this takes place the term
uterus may be applied.

The testes offer no difficulties of terminology. *Vasa efferentia* may
be applied to cases such as Polyclads, where a fine duct passes from
each testis-follicle, and *vasa deferentia* to the paired canals formed by
their union. These canals usually open into a *vesicula seminalis.*
Accessory male glands are very commonly present, and possess fairly
uniform histological characters. Hence the terms *granule-gland,
granule-duct, vesicula granulorum.* The duct through which the male
products reach the exterior is the *ductus ejaculatorius,* and any
chitinous investment round it may be called a *copulatory organ.*

As regards the authors' names appended to the species, I have
endeavoured to follow the British Association rules. Von Graff, in
his "Monograph,' has employed the name of that author who first
used the definitive combination of genus and species. Thus *Vortex
balticus,* M. S. Schultze, becomes *Provortex balticus,* v. Graff, whereas
I write it *Provortex balticus* (M. S. Schultze).

The terminations of generic names have not hitherto been formed in an uniform way. Von Graff changed the terminations of all the older generic names, such as Acmostomum, to Acmostoma, whereas Lang retains the -um form. In the present paper I have followed these authors. It would, however, seem advisable in the future to adopt either one termination or the other, right through the group. Spengel's* Latinised form of ' Alloiocœla" (Allœocœla) is adopted.

<h2 style="text-align:center">II.—Systematic.</h2>

<h1 style="text-align:center">TURBELLARIA.</h1>

<h3 style="text-align:center">Sub-order 1.—RHABDOCŒLIDA.</h3>

<h3 style="text-align:center">A. ACŒLA.</h3>

A digestive cavity absent. Mouth ventral, leading indirectly through the pharynx into the parenchyma. A " frontal gland " and otolith present. Nervous system consisting of a brain and peripheral nervous sheath. Hermaphrodite. Testes follicular, rarely compact.

<h3 style="text-align:center">Family PROPORID.E.</h3>

Acœla with a common genital pore.

Genus 1.—PROPORUS, v. Graff
(= SCHIZOPRORA, Schmidt, 28).

Proporidœ without spermotheca.†

1. PROPORUS VENENOSUS (O. Schmidt, 28).

Length 1 mm. *Body* elongate, cylindrical, rounded at both ends. *Colour* bright yellow, due to diffused granular pigment. *Epidermis* ciliated, containing numerous *rhabdites*, some free, some grouped in formative cells. These pyriform groups are specially abundant towards the hinder end, giving a spinous appearance to the surface. The *mouth* lies just beneath the anterior end ; circular at rest, it becomes slit-like during movement. The *pharynx* is a direct invagination of the anterior end (v. Graff, 'Acœla,' pl. x, fig. 5). It is

* 'Götting. gelehrte Anzeigen,' March 1st, 1884, p. 183, note.

† The definitions of families, sub-families, and genera are taken from v. Graff (53).

long (one-fourth the length of the body), cylindrical, and muscular. The *otolith* has a distinct central portion, and is radially striated. The two *eyes* are large, and provided each with a lens. The *common genital pore* lies at the hinder end of the body. It leads into a ciliated, muscular, narrow *atrium*, which is an invagination of the hinder extremity. *Testes* rounded, scattered. *Vesicula seminalis* spherical, opening into the pyriform muscular penis enclosed in its sheath. These organs lie at the front end of the *genital atrium*. The *spermatozoa* when mature are broad and thick, tapering at either end, the head filament shorter than the tail. *Ova* occur in two lateral rows, sometimes those of one side being more developed than those of the other (v. Graff).

HABITAT.—This active yellow form is not uncommon between tide-marks in Plymouth Sound (F. W. G.). This is, I believe, the first record of its occurrence north of the Mediterranean.

DISTRIBUTION.—Trieste, Naples, Messina (v. Graff), Lesina (Schmidt, 28), Sebastopol (Uljanin, 41).

Genus 2.—MONOPORUS, v. Graff (56)

(= PROPORUS, Schmidt, 28).

Proporidæ with spermotheca.

2. MONOPORUS RUBROPUNCTATUS (O. Schmidt, 28).

Length 1 mm. *Body* ellipsoidal with rounded extremities. *Colour* white, the central parenchyma brown, owing to the presence of coloured vacuoles and food-particles. The *epidermis* contains clavate *rhabdites*, the thickened outer ends of which project beyond the surface. The *mouth* is mid-ventral, leading into a very short simple *pharynx*; the "terminal mouth" of earlier descriptions being, according to v. Graff (56, p. 57), the opening of the "frontal organ." The *eyes* lie right and left close to the anterior end. They are placed in the epidermis, and consist of polygonal pigment masses of a brilliant carmine colour. In his recent accounts of the " Acœla " v. Graff gives some interesting additions to our knowledge of the genital organs. The *genital pore* is single. The *testes* are compact as in no other Acœlous form. The *vasa deferentia* are merely their posterior prolongations. The *oviducts* are continuous with the epithelial lining of the

. . . unite to form a *vagina*. Opening into this . . . . a
. . . . duct of which is slightly chitinised.

. . . . . littoral algæ, Plymouth Sound (F. W. G.).

DISTRIBUTION.—Naples, Trieste, Dalmatian Islands (v. Graff, 55),
Lesina (Schmidt, 28).

APHANOSTOMID.E.

*Acela with two genital apertures, the . . . . . . of the
. . . . A spermotheca present.*

Genus 3.—APHANOSTOMA, Oersted (21).

Spermotheca unarmed.

3. APHANOSTOMA DIVERSICOLOR, Oersted (21).

Length ·75—1 mm. *Body* somewhat fusiform, tapering gradually
forward from the posterior third, more rapidly backwards to the hinder
end. The *colour* (which is variable in amount and intensity) is due to
violet pigment-cells and yellow vacuoles in the *parenchyma*. These are
usually disposed in a characteristic way. The yellow pigment is
present at the anterior end, and extends a short distance backwards
on each side, enclosing the violet pigment which occupies the median
part of the body as far back as the limit of the anterior third.
A small patch also occurs at the extreme hinder extremity. The
violet . . . . . cells are capable of altering their shape according to
the . . . of contraction of the body. Their most characteristic form
is . . . of a **U**, the curved portion being much thicker than the long
. . . .

. . . . . . is almost mid-ventral. It leads into a funnel-shaped
. . . . . . Two *genital apertures* are present. The male pore lies a
. . . . in front of the hinder end, the female pore still further
forward. The conical *penis* encloses a *vesicula seminalis* in its proximal
part. The *ovaries* extend throughout the greater part of the body.
A . . . . possessing intrinsic and extrinsic muscles, opens into
the . . . . . . The epithelium of its duct (according to v. Graff) . . . . . .
a cuticle.

HABITAT.—Among stones and seaweed at the base of the littoral zone, Plymouth, Port Erin, Isle of Man (F. W. G.) ; Millport (v. Graff).

DISTRIBUTION.—Naples, Trieste, Roscoff (v. Graff) ; Denmark (Oersted, 21); in colonies among Laminaria and Fucus, a few feet below the surface, Bergen (Jensen).

4. APHANOSTOMA ELEGANS, Jensen.

Length ·75 mm. *Body* colourless, with a lobate dark green spot in the centre due to coloured parenchymatous vacuoles. Form broadly rounded in front, tapering gently posteriorily. *Eyes* absent. The male *genital aperture* lies a short distance in front of the hinder end : the *female pore* close behind the green spot. According to Jensen the *spermatozoa* are long, filiform, thicker and spirally twisted anteriorly.

HABITAT.—Among Ulva, between tide-marks, Plymouth (F. W. G.).

DISTRIBUTION.—Alvœrström and Bergen (Jensen, 49).

Jensen's description reads, "Corpus utraque extremitate rotundatum in anteriore parte latius, retrorsum sensim angustius." My specimen was certainly during active motion broader in front than behind, approaching Jensen's figure of *A. rhomboides* (49, pl. i, fig. 1).

Aphanostoma is a distinctly northern genus. The coast of Denmark and the western shores of Norway and Greenland have furnished the bulk of the existing records. The presence of *A. elegans* at Plymouth suggests that further search will reveal localities for the remaining species on our coasts.

Genus 4. CONVOLUTA, Oersted (16), 1844.

Aphanostomidœ with a broad, flat body, the margins of which are in some forms capable of being flexed ventrally. Spermotheca with a chitinous mouth-piece.

5. CONVOLUTA SALIENS, v. Graff, 1891.

1882. CYRTOMORPHA SALIENS, v. Graff, (53).

Length 1 mm. *Body* elliptical, the dorsal surface convex, ventral surface flat. *Colour* is absent except in the centre, where it is due to brown food-particles among the parenchyma. Locomotion is effected

thus :—From the anterior end backwards for one-third of its length
the margins of the body are capable of being gradually extended
outwards, so that the greatest width of these animals when in motion
is a short distance in front of the centre of the body. These lappets
are then flapped inwards and downwards, the animal at the following
instant leaping forward. When, however, contraction occurs, it is no
longer possible to define the lappets. The anterior end is at a much
lower level than the rest of the dorsal surface. As the change of level
is abrupt the front end appears snout-like, especially when seen from
the side. This snout is moved from side to side in a sensitive manner.
Short *cilia* and slightly irregular *rhabdites* are present in the *epidermis*,
and are disposed in alternate longitudinal rows, which converge
anteriorly towards the opening of "*frontal gland.*" *Eyes* are absent.
The *otolith* is concavo-convex, with a central "nucleus." The two
genital pores are posterior, the female in front of the male. Opening
to the exterior through the former is the *spermotheca*. The curved
penis receives the contents of the *seminal vesicle*.

HABITAT.—In tide-pools, Millport (v. Graff). Two specimens
among Zostera and Corallina, Plymouth (F. W. G.).

6. CONVOLUTA PARADOXA, Oersted (16), 1884.

1777. PLANARIA CONVOLUTA, *O. F. Müller* (4).
1844. CONVOLUTA PARADOXA, *Oersted* (16).
1845. PLANARIA MACROCEPHALA, *Johnston* (20).
1853.   ,,   HAUSTRUM, *Dalyell* (29).
1853. CONVOLUTA PARADOXA, *Gosse* (30).
1861.   ,,   ,,   *Claparède* (35).
1865.   ,,   ,,   *Johnston* (38).
1866.   ,,   ,,   *Lankester* (39).

Length 1—3·5 mm. Schmidt (28) found specimens up to 9 mm. in
length in the Faroe Islands and other northern localities ; v. Graff
records equally large examples from Millport. Specimens from more
southerly places are usually much smaller, and it might therefore
appear that this species attains a larger size in the northern than in
the southern seas. Claparède, however (35), working on the north-
west coast of Skye, examined a large number of specimens, none of
which exceeded 2 mm. in length ; while recently (55) v. Graff has
found "giant specimens" at Rosscoff. It seems more likely,

Claparède suggested, that this species attains a considerable size before the reproductive organs begin to develop. Thus he found individuals 1·5—2 mm. long without a trace of gonads, and I have myself observed the same thing.

The form of the body changes with different states of contraction and expansion. When freely swimming the form is that of (53) pl. xi, fig. 15, the sides flexed ventrally, almost touching one another except in front, where they diverge. The hinder end is produced into a finely-pointed tail. The anterior end is truncate, the angles being frequently more or less produced. It is a most actively sensitive animal, especially during creeping movements, when the "head" is converted into a funnel-shaped structure which explores the surroundings.

The general *colour* varies from greenish-brown to a warm chestnut-brown, which is the usual tint. The anterior margin is paler than the rest. The brown pigment is deposited in the cell of a "symbiotic alga," the nature of which has not been thoroughly investigated. Transverse, narrow white bars, 1—2 in number (v. Graff has observed three, and Claparède [35, pl. vi, fig. 2] figures four such bands), are present in large individuals (1·75 mm. and upwards). They are the expression of a large number of very small irregular granules, insoluble in acids (v. Graff). Claparède has suggested that these bars may be a "caractère sénile." This view is supported by v. Graff, since he finds that the bars become more and more distinct with the increased size and age of the animal. The *epidermis* contains *flagella, rhabdites*, and *pigment*. The latter forms elongate masses of rod-like granules. The *mouth*, which is ventral and subcentral, leads into a very short pharynx. The *eyes* are constantly present in this species. They are red pigmented bodies, and occur right and left of the otolith. *Poison-organs* ("Gift-Organe") have been discovered by v. Graff (44) in this and other species of *Convoluta* (*C. grœnlandica, cinerea, flavibacillum, bimaculata*). They consist in *C. paradoxa* of a pair of pyriform, transparent, muscular vesicles, provided with hollow chitinous tips, and are placed at the margin of the body in such a way that when this is flexed ventrally the tips are directed towards the mouth. The contents consist of small refractive granules. At each contraction of the muscular wall the tip is moved forwards, at the same time discharging some of its contents. This oral pair of poison-organs is constantly present in sexually mature individuals. When the male products ripen (this species is a distinctly protandrous hermaphrodite) two other pairs (according to v. Graff) arise close to the male genital pore. They differ from the oral pair in three points— their time of appearance, their variability (the hinder pair may be absent), and their disappearance after the shedding of the male products (see v. Graff, 44, p. 61). The *female genital pore* lies slightly in front of the centre, the male pore halfway between it and the hinder end. The *testes* are dorsal in position. The *penis* is a narrow, cylindrical, muscular tube opening into the short *genital atrium*, surrounded by radiating accessory glands. The *spermatozoa* are long (·22 mm.), and consist of a finely granular central portion and hyaline borders, absent, however, on the distinct "tail." The *ova* are placed ventrally, and when mature are ovoid, ·07 mm. diameter, containing much yellow food-yolk, and surrounded by a delicate membrane. Thirty to forty fertilised eggs may be present at one time in a single example. A *spermotheca* is present. Its neck is produced forwards into a funnel-shaped expansion opening into the female atrium and backwards into the swollen basal portion containing spermatozoa. Round the neck chitinous plates are arranged one over another, the margins of which are thin and colourless, forming the "mouth-piece." Young specimens of *Convoluta paradoxa* differ from the adults, to which the foregoing description applies, chiefly in the absence of reflexed marginal lappets, and the small number (five to seven) of symbiotic algæ present.

HABITAT.—A littoral species, occurring among seaweeds in tide-pools all round our coast. Berwick Bay (Johnston) ; Firth of Forth (Dalyell) ; among Ceramiæ, Weymouth (Gosse) ; Skye (Claparède) ; Guernsey (Lankester) ; St. Andrews (McIntosh) ; Millport (v. Graff) ; Plymouth, Port Erin, Isle of Man (F. W. G.).

DISTRIBUTION.—Mediterranean, Adriatic and Black Seas, North Sea, North Atlantic, coast of Denmark.

7. CONVOLUTA FLAVIBACILLUM, Jensen.

Length 2—3 mm. *Body* stout, oval, pointed behind, dorsal surface convex, ventral surface flat. The *margins* are produced into thin

N

lamellæ, flexed ventrally. *Colour* yellow, due to the prevalence of
pigmented yellowish-green rods over irregular brown granules. *Epidermis, flagella,* and *rhabdites* are similar to those of *C. paradoxa.*
Eyes two, reddish, placed on each side of the *otolith*. *Female genital
pore* situated at one-third the length of the body from the hinder end,
halfway between the latter and the pore is the *male aperture*. The
chitinous mouth-piece of the *spermotheca* is cylindrical, and composed
of a number of perforated plates placed one upon another. The
spermatozoa are elongate, tapering at each end, with a granular central
portion and clear lateral membranes. *Penis* cylindrical, muscular.

HABITAT.—Among Laminaria, &c., Millport (v. Graff); among sand
at base of *Corallina officinalis* in tide-pools, Port Erin, Isle of Man;
Plymouth (F. W. G.).

DISTRIBUTION.—Bergen (Jensen).

The stout form and yellow colour sufficiently serve to distinguish
this species, which, in its anatomy, closely resembles *Convoluta
paradoxa.*

B. *RHABDOCŒLA.*

*Gut and parenchyma distinct. A spacious body-cavity usually present.
Nervous and excretory organs present. Gonads hermaphrodite (except
Microstoma and possibly Stenostoma). Testes usually compact, and
ovaries, germ-yolk-gland, or separate vitellaria and germaria may be
present; in all cases surrounded by a tunica propria. Pharynx
present, variable. Otolith usually absent.*

Family MICROSTOMIDÆ.

*Rhabdocœla with sexual and asexual reproduction. Female accessory
apparatus absent. Pharynx simple.*

Genus 5.—MICROSTOMA, O. Schmidt, 1848 ('Die rhabd. Turbellarien
d. süssen Wassers').

*Microstomidæ with separate sexes and compact testes. Body uniformly
ciliated, provided with "ciliated grooves" and a pre-œsophageal
gut-cæcum.*

8. MICROSTOMA GRŒNLANDICUM, Levinsen (51).

Length 1·75 mm. *Body* composed of about eight zooids, in colour resembling *M. lineare*. *Eyes* absent. *Ciliated grooves* small. *Rhabdites* well developed anteriorly, more sparsely behind. No sexual organs were observed.

HABITAT.—Among Ulva, Plymouth Sound (F. W. G.); Millport (v. Graff).

DISTRIBUTION.—Egedesminde, Greenland (Levinsen).

I place the Plymouth specimen here, for although it does not agree exactly with Levinsen's original description (the red spot at the anterior end is wanting), it does agree with a form described by v. Graff from Millport, and placed under this species.

Genus 6.—ALAURINA, Busch (26).

Hermaphrodite Microstomidæ, with tactile anterior " proboscis." Usually with posterior setæ ; paired lateral ones sometimes present.

9. ALAURINA CLAPAREDII, v. Graff.

Length ·3 mm. Anterior end modified to form a tactile proboscis provided with numerous circular folds. The base of the proboscis is marked off by a tuft of cilia on each side. A single posterior group of setæ at the hinder end. The posterior sixth of the body is marked off transversely (probably an indication of fission).

HABITAT.—Coast of Skye (Claparède).

Claparède (35) described this form as a rhabdocœle larva, but it was referred to *Alaurina* by Metschnikoff (37). It is probable that this genus is not uncommon on our coasts.

Family MESOSTOMIDÆ.

Rhabdocœla with one or two genital apertures ; yolk-glands and germaria distinct or united. Testes compact. Female accessory organs present. Pharynx ventral, rosulate (for the term "rosulate" see v. Graff, ' Monogr.,' p. 80).

Subfamily PROMESOSTOMINÆ.

Genus 7.—PROMESOSTOMA, v. Graff.

Mesostomidæ with two germaria and separate yolk-glands, and a common genital aperture. Female accessory organs absent. Testes small.

10. PROMESOSTOMA MARMORATUM (Schultze [27]). Pl. X, fig. 10;
Pl. XI, fig. 16.

Length ·5—1·5 mm. *Body* elongate, cylindrical, broadly rounded in front, truncate behind. Anterior end used as a tactile organ. *Colour* very variable. The epidermis which furnishes the ground colour is colourless, bright yellow, or yellowish-red. Black reticular pigment is almost constantly present as a small patch between the eyes ; elsewhere to a variable extent, and may be entirely absent. The epidermis on the inner side of the eyes contains immense numbers of rhabdites (fifteen to twenty in a single mother cell), forming two well-defined tracks. Elsewhere they are few in number. The posterior end is provided with *adhesive papillæ*. The *pharynx* lies just behind the centre of the body. The commencement of the gut (as v. Graff observed) is marked by active *flagellæ*. The *genital aperture*, provided with a muscular lip, is placed behind the pharynx. *Testes* two, oval. The connection between the two paired *vasa deferentia* has not actually been traced. From their point of union (behind the genital aperture) the single duct runs forward and expands into a *vesicula seminalis*, which is partly filled with sperm, partly with the *granule-secretion*. Both these products are conveyed to the exterior by a very curious *copulatory organ*. The most typical form which this chitinised ejaculatory duct assumes is that of a bishop's crosier, Pl. XI, fig. 16. The variations both in the number and form of the coils, and also in the form of the tip of the organ (straight, curved, forked), are great. The limits of variation in different directions would become of specific value if intermediate forms were not known to occur. The *germaria* and *yolk-glands* are paired and lateral. The *egg-capsules* are stalked.

HABITAT.—This very active littoral species has been found at Millport (v. Graff) ; Kilmore, Skye (Claparède) ; Port Erin, Isle of Man (F. W. G.) ; Plymouth (F. W. G.). At Millport and Plymouth forms with "long" and "short" copulatory organs occur; at Trieste and Naples v. Graff found only forms with short ones. At Plymouth most of the specimens had little or no pigment.

DISTRIBUTION.—Naples, Messina, Black Sea, Baltic, North Atlantic.

11. PROMESOSTOMA OVOIDEUM (O. Schmidt, 28).

Length ·5 mm. *Body* oval. The dense black colour is due to reticular parenchymatous pigment, which forms a thick mesh-work round the internal organs. The *epidermis* contains large numbers of rhabdites, especially developed along the inner side of the eyes, which are reniform, and provided with a lens. v. Graff has observed trembling movements of the eyes. I have described them as seen in *Pr. solea*. *Pharynx* in the posterior third of the body. Behind it lies the *penis*, which is pyriform, its upper part filled with *spermatozoa*. The duct is chitinous.

HABITAT.—This species is found rarely in 5—15 fms. Plymouth Sound (F. W. G.).

DISTRIBUTION. — Messina, Naples (v. Graff), Lesina (Schmidt), Egedesminde, Greenland (Levinsen).

12. PROMESOSTOMA SOLEA (O. Schmidt, 32).

Differs from the preceding species in two points. The reticular pigment is less dense, and the pigment-cup of the eye sends a hooked process over the outer surface of the lens. The latter point alone seems to me to be constant. The amount of reticular pigment varies greatly. Seen from the dorsal surface the eye has an appearance similar to a miniature pan or tobacco pipe, the bowl being represented by the pigment cup, and the stem or handle by the strip of pigment running over the lens. The vibratory movement of the eye is performed in the following way. Suppose the pan or pipe to vibrate through a small angle in its plane of symmetry, in such a way that the plane is horizontal, the bowl moving forwards and then backwards. The actual vibrations of the eye are of this kind. Apparently one eye commences, performs five or six vibrations in a second, and then stops; the other eye begins, and so on. I am not certain, however, that the movements are alternate for any length of time. Of the mechanism I am ignorant.

HABITAT.—This is a typical deep-water (8—20 fms.) Turbellarian. Only once have I taken it between tide-marks. When dredge-material is placed in sea water, dark oval specks (the present species) are often seen swimming actively at the surface. In a few hours they descend, and reappear only when the water begins to foul. Plymouth (F. W. G.).

DISTRIBUTION.—Naples (Schmidt and v. Graff), Messina (Graff), Sebastopol (Uljanin).

13. PROMESOSTOMA LENTICULATUM (Schmidt [28]).   Pl. X, fig. 6 ; Pl. XI, figs. 13, 17.

Length ·65—·7 mm. (*i.e.* half that of Schmidt's specimen). *Body* broadly truncate and slightly convex in front, the antero-lateral margins produced slightly outwards. Behind these it becomes narrower, forming a " neck ; " it then widens towards the middle, diminishing again to the posterior end. The general shape is, in fact, similar to Jensenia (see Jensen [49], pl. iii, figs. 1, 2), but more elongate. *Colour* to the naked eye scarlet ; this is due to the contents of the extensive gut. *Movements* extremely active. *Epidermis* very transparent. Rhabdites few, scattered. *Pharynx* placed slightly in front of the middle of the ventral surface. Intestine large, corresponding to the shape of the body. The *eyes* are provided with a large conspicuous lens, which is easily detached from the pigment-cup. The *genital aperture* is a short distance behind the pharynx. The *testes* have the usual relations, and lead at their posterior ends into the *vasa deferentia*, which unite to open into the base of a most remarkable *copulatory organ*. This is cylindrical at its proximal end, provided distally with a series of triangular chitinous plates ranged round the terminal slit. The whole resembles the tool known as a " rose-bit " or " counter-sink," and used for embedding the heads of screws in wood or metal. The base of this organ is divided into spermatic and granule portions. The *germaria* are placed posteriorly. On one side the *germarium* was normal ; on the other it was composed of lenticular masses, with difficulty separable optically from one another. The paired *yolk-glands* occupy the greater part of the sides of the body.

HABITAT.—Two specimens from a tide-pool among corallines, Port Erin, Isle of Man (F. W. G.).

DISTRIBUTION.—Faroe Islands (Schmidt).

This species has hitherto only been seen by Schmidt, who gave no account of the genital organs. He described the form and colour, the position of the pharynx, and the eyes. Excepting the difference in size (Schmidt's specimens measured 1·5 mm.) I have no reason for doubting the identity of my specimens with his.

As regards the systematic position of this species, my observations are not perfectly conclusive. It is possible that further investigations may prove that the organ R. S. is a receptaculum seminis, and B. C. a bursa copulatrix, the transverse markings I have noted being the subspiral muscles of Jensen. At present, however, from the nature of the protoplasm, I believe it to be an ovary, while the presence of a single genital opening is evidence for a position in the genus Promesostoma, where v. Graff has already placed it doubtfully.*

14. PROMESOSTOMA AGILE (Levinsen).   Pl. XI, fig. 14.

Length ·5 mm. *Body* oval, rounded posteriorly, tapering forwards in front of the hinder third. *Colour* light red. *Movements* very active. *Eyes* placed close together, triangular, the apex being formed by the pigment-cup, directed inwards and backwards, the lens outwards and forwards. *Pharynx* subcentral. *Intestine* reddish, occupying the space between the pharynx and the lateral yolk-glands. The *genital aperture* is placed about halfway between the pharynx and the posterior end. The *testes* are two oval sacs behind the pharynx; they communicate by short ducts with the *penis*, the base of which is spherical, and contains the secretion of a *granule-gland*, while its distal portion is produced into a long, narrow, slightly curved *copulatory organ*. A pair of *ovaries* are placed at the posterior end of the body, their ducts running forwards to the genital aperture. The *yolk-glands* are lateral uniting behind the brain. A small muscular sac placed behind the penis appears to be a *receptaculum granulorum*.

HABITAT.—Two specimens among littoral weeds, Plymouth (F. W. G.).

* A second genital pore might have easily been overlooked. If further examination should demonstrate a second aperture, the species would have to be transferred to the genus Byrsophlebs.

DISTRIBUTION.—West coast of Greenland (Levinsen).

Levinsen's description (according to v. Graff) left this form indeterminate. The two pyriform bodies that he called "Samenblasen," and which v. Graff suggested might be local swellings of the vasa deferentia, I take to be the true testes. They have all the structure of such an organ, but occupy a more posterior position than usual. The ova did not appear to Levinsen to be distinguishable into two fairly distinct ovaries as in my specimens.

Subfamily BYRSOPHLEBINÆ.

Mesostomidæ with two genital apertures, the male in front of the female. Germarium single. Vitellaria distinct. Accessory organs absent or present. Testes small, rounded.

Genus 8.—BYRSOPHLEBS, Jensen (49.)
(*With the diagnosis of the subfamily.*)

15. BYRSOPHLEBS GRAFFI, Jensen.

Length ·45 mm. Body cylindrical, tapering gradually posteriorly. Colourless, the gut brownish-yellow. *Pharynx* central, male *genital aperture* immediately behind it, female aperture close to the hinder end. *Penis* composed of a proximal cylindrical portion, strengthened by spiral and longitudinal muscles, and a distal chitinous funnel-shaped duct. The terminal aperture of this duct is provided with a short triangular projection on one side. Opening through the female genital aperture is a *receptaculum seminis* placed at the base of the ovary, and a muscular *bursa copulatrix* at its side. *Yolk-glands* unbranched, lateral. (For further account with figures see Jensen (49), pl. ii, figs. 8—12.)

HABITAT.—Among algæ, Drake's Island, Plymouth Sound (F. W. G.); among Ulva and Fucus, Millport (v. Graff.)

DISTRIBUTION.—Bergen and Sund, West Norway (Jensen).

16. BYRSOPHLEBS INTERMEDIA, v. Graff.

Resembles the former species in most characters, differing in the arrangement of the genital organs. The *penis* is elongate, cylindrical, the spiral muscular fibres surrounding the united vasa deferentia.

The chitinous portion is narrow, funnel-shaped, its tip partly surrounded by a curved process, similar to the "spur" in *Macrorhynchus Naegelii*. The *yolk-glands* are branched.

HABITAT.—In tide-pools. Port Erin, Isle of Man (F. W. G.); Millport (v. Graff).

Subfamily PROXENETINÆ.

Mesostomidæ with one common genital aperture and two germ-yolk-glands. Spermotheca provided at its blind end with chitinous appendages. Testes small, rounded. Copulatory organ complicated.

Genus 9.—PROXENETES, Jensen (49).

17. PROXENETES FLABELLIFER, Jensen.

Length 1—1·5 mm. *Body* cylindrical, abruptly rounded in front, narrowing posteriorly to a short "tail," colourless. *Flagella* are everywhere present between the cilia. Long, sharply-pointed rods are present in great numbers in the epidermis, forming two well-defined tracks between the eyes and supplying the anterior end. Smaller rods occur over the rest of the surface. The hinder end is provided with adhesive cells. *Pharynx* large, placed in the posterior third. Between it and the hinder end is the *genital aperture*. The large *testes* occupy the middle of the sides of the body. The *vasa deferentia* are swollen just before uniting at the base of the retort-shaped *penis* which receives the secretion of *granule-glands*. The *copulatory organ* is a complicated mechanism of chitinous pieces separating the granule-secretion from the spermatozoa. The nutrient part of the *germ-yoke-gland* extends along the sides of the body; the ovarian portion develops behind the pharynx. The *two oviducts* unite, and the common duct runs to the genital pore. Opening into the atrium, close to the pore, is the large *spermotheca*, which is directed forwards towards the pharynx, and then bends back upon itself. Its blind end receives chitinous ducts conveying granule-secretion. The point of connection with the genital atrium is armed with five to six triangular chitinous teeth, freely hinged at their bases. A further account of these structures will be found in v. Graff (pp. 277—279) and Jensen (49). The above are points I have verified myself.

HABITAT.—Several feet below surface and in tide-pools. Millport (v. Graff); Plymouth (F. W. G.); common among *Ptilota plumosa* and other red and green algæ, low spring tide, Port Erin, Isle of Man (F. W. G.)

DISTRIBUTION.—West coast of Norway (Jensen).

18. PROXENETES COCHLEAR, v. Graff.

The four or five chitinous ducts of an accessory gland, which open into the blind end of the bursa in *P. flabellifer*, are, in this species, reduced to a single spiral one; the triangular "teeth," arming the duct of the bursa, being here represented by a series of small chitinous processes. The copulatory organ consists of three spoon-shaped pieces fitting into one another. The spaces between them constitute the passages for the seminal and granule fluids.

HABITAT.—Millport, one specimen (v. Graff).

Subfamily EUMESOSTOMINÆ.

Mesostomidæ with one common genital pore, a single germarium, two vitellaria and accessory organs. Testes long. Excretory vessels opening into the pharyngeal sheath.

Genus 10. MESOSTOMA, Dugès.

Without otolith. Copulatory organ traversed throughout its entire length by the ducts of male secretions.

19. ? MESOSTOMA NEAPOLITANUM, v. Graff.

Length ·5 mm. *Body* flattened behind, bluntly pointed in front, white. Between the *eyes* are two tracks formed by masses of large curved *rhabdites*. Smaller ones are present elsewhere. *Pharynx* small, in front of the centre. *Intestine* large, filling up the greater part of the body. *Eyes* two, with lenses. *Genital* pore, close to the hinder end. *Testes* lateral. The *penis* bears a funnel-shaped terminal portion which receives the contents of the *vesicula granulorum*, and in front of this lies the semi-lunar *vesicula seminalis*. The atrium did not appear to be so large as in v. Graff's specimen.

HABITAT.—Among Fucus, Plymouth Breakwater (F. W. G.).

DISTRIBUTION.—Naples (v. Graff).

I append a query to this species, since my observations were made on one specimen, and agreed more closely (but not entirely) with *M. neapolitanum* than with any other species. More specimens of this marine Mesostoma are greatly needed, as its position is not thoroughly defined.

Family PROBOSCID.E.

Rhabdocœla with tactile proboscis ; one or two genital pores. Germaria and vitellaria distinct. Testes compact. A spermotheca present. Mouth ventral. Gut discontinuous in the adult, owing to the development of the gonads. Copulatory organ complicated, chitinous (v. Graff, p. 315).

Subfamily PSEUDORHYNCHIN.E.

Proboscis without sheath or muscular cone. Retractors represented by short muscular bundles. Pharynx rosulate. One genital pore. Two germaria. Yolk-glands reticular. Testes paired, rounded.

Genus 11.—PSEUDORHYNCHUS, v. Graff.

20. PSEUDORHYNCHUS BIFIDUS (McIntosh).

1875   MESOSTOMUM BIFIDUM, *McIntosh* (45).

Length 1·3 mm. *Body* convex dorsally, flat on the ventral surface, slightly expanded towards the posterior end, which is bifid. The conical anterior extremity (proboscis) is devoid of cilia. *Colour* pale orange with darker spots, the proboscis colourless. *Rhabdites* are well developed and of three kinds (Jensen)—straight, ovoid, and needle-shaped. Strong adhesive papillæ occur at the bifid hinder end. *Pharynx* subcentral. The *genital pore* is placed behind the middle. *Testes* small, rounded. *Vasa deferentia* unite in a *vesicula*, which opens, along with the ducts of "granule-glands," into the proximal swollen parts of the penis. The distal portion is muscular, and contains the *copulatory organ*. The latter is a conical chitinous tube, the outer wall of which is produced into a series of spiral ridges running from the base to the apex in a screw-like manner. The usual

direction appears to be " right-handed." von Graff notes an interesting
"left-handed" variety. The *germaria* are large lateral oval sacs, placed
opposite the centre, and directed forwards. The *spermotheca* is
finger-shaped with muscular walls. It contains the granular secretion
of a large number of glands.

HABITAT.—On half decayed Laminaria and Ulva, Millport (v. Graff);
under stones between tide-marks, St. Andrews (McIntosh, 45); among
decaying Ceramia, &c., Port Erin, Isle of Man (F. W. G.).

DISTRIBUTION.—Egedesminde, West Greenland (Leviusen), Faroe
Islands (Schmidt), Bergen (Jensen).

Sub-family ACRORHYNCHINÆ.

*Proboscis at the anterior end provided with a sheath opening in front, with
muscular cone and four long retractors. Pharynx rosulate. Yolk-
glands reticular* (v. Graff [53], p. 318).

Genus 12.—ACRORHYNCHUS, v. Graff.

*Acrorhynchinæ with a common genital pore. Two germaria and elongate
testes. Vesiculæ seminales and granulorum distinct, but enclosed in
a common muscular penial sheath. The copulatory organ transmits
both secretions* (v. Graff, 'Monogr.,' p. 319).

21. ACRORHYNCHUS CALEDONICUS (Claparède).

1861. PROSTOMUM CALEDONICUM, *Claparède.* (35).

Length 1—2 mm. *Body* bluntly pointed in front, gradually
widening behind, and rounded posteriorly. *Colour*, proboscis white,
the rest of the body greyish brown with light areas, indicating the
positions of the more bulky internal organs. *Rhabdites* very small,
numerous, evenly distributed. *Proboscis* well developed (for an
account of its structure see v. Graff, 'Monographie,' pp. 119—124).
Two very long retractor muscles extend from its base to the posterior
end of the body. *Pharynx* before the middle, with a marginal "seam."
According to von Graff, the base of the extended proboscis is enclosed
in a nervous commissure proceeding from the brain. *Eyes* two,
provided with a lens. The *genital pore* is placed just behind the centre.

Testes lie at the sides of the pharynx. The *vasa deferentia* unite in a *sperm-vesicle*, close to which is the *granule vesicle*. The long muscular cylindrical penis receives both secretions, and is enclosed in an inner chitinous sheath and an outer muscular one. The armature is in the form of hooks with rounded basal ends. The hooks vary in size and shape, and are disposed as in v. Graff's pl. x, fig. 17, of his 'Monographie.' The *genital aperture* is surrounded by numerous glands. The *germaria* lie at the sides of the *penis*. The reticular *yolk-glands* are of considerable extent.

HABITAT.—In tide-pools, and among Fucus and Laminaria below the surface of the sea. Kilmore, Skye (Claparède) ; Millport (v. Graff) ; Plymouth ; Port Erin, Isle of Man (F. W. G.).

DISTRIBUTION.—Bergen (Jensen), Heligoland (Metschnikoff).

The well-developed musculature of these forms enables them to resist compression without rupture. If a mature individual be treated in this way the various parts of the genital apparatus stand out with diagrammatic clearness.

This species, closely similar to the next in external appearance, can be distinguished by the presence of the common muscular envelope round the vesiculæ seminales and granulorum.

Genus 13.—MACRORHYNCHUS, v. Graff.

Acrorhynchinæ with a common genital aperture, two germaria, and paired elongate testes. Vesiculæ seminales and granulorum distinct, the duct of the latter with a special chitinous tube ('Monogr.,' p. 321).

Von Graff divides this genus into two subdivisions:—i. *Typici.*—Those in which a poison-dart (Gift-Stachel) is absent ; duct of the vesicula seminalis without chitinous armature. ii.—*Venenosi.*—Those provided with a poison-dart ; a chitinous investment for both the duct of the seminal and granule-vesicles.

i. TYPICI.

22. MACRORHYNCHUS NAEGELII (Kölliker, 17). Pl. X, fig. 5 ; Pl. XI, fig. 15.

Length 2·2 mm. in my largest specimens ; v. Graff states that he has not seen larger ones at Millport, while Mediterranean examples range up to 4 mm. *Body* similar to *Acrorhynchus* in shape, cylindrical, bluntly pointed in front, broader behind. *Colour* variable. *Proboscis* usually white, the rest of the body dusky brown, through which the testes, ovaries, and penis can be indistinctly perceived with a hand-lens. Claparède (36) has described a variety from St. Vaaste-la-Hogue, which was white with a median dorsal yellow streak. A similar variety occurred at Plymouth (fig. 5). The colour was brown; at the base of the white proboscis and in the middle line was a yellow spot, visible in all positions of the animal; proceeding from this backwards along the mid-dorsal line was a bright yellow streak. These colours appear to be due to pigment in the parenchyma. Small rhabdites are plentifully developed in the *epidermis*, and the epithelium of the proboscis contains oval, highly refractive bodies, "Nematocysten entsprechenden Gebilde" (v. Graff, 'Monogr.,' p. 323). The histology and relations of the proboscis have been fully elucidated by v. Graff. The *brain*, bearing two eyes provided with lenses, is placed behind the proboscis, and behind this again is the spherical *pharynx*. The common *genital pore* lies behind the centre of the body. It leads into a very extensive *atrium*, the chitinous lining of which extends into the common passage receiving an anterior male subdivision and a posterior female portion. The *testes* occur at the sides of the body. The *vasa deferentia* unite with the contents of an accessory gland. The secretion of the *granule-gland*, however, is enclosed in a special sac, the lower part of which is chitinous. Its free edge is usually provided with a curved "spur," of variable form. This may be absent. A remarkable variety is figured in Pl. XI. fig. 15. The *ovaries*, and between them the large *spermotheca*, lie posteriorly. One egg-capsule was present in many individuals at Plymouth in September. In November no adults could be found. They had probably died off.

HABITAT.—Similar to *Acrorhynchus caledonicus*. The two are, however, only occasionally found together. Plymouth Sound (F. W. G.); Millport (v. Graff, "Ich selbst habe in Millport niemals grössere Exemplare gesehen als Claparède's," p. 323, 'Monogr.').

DISTRIBUTION.—This is the commonest Rhabdocœle at Naples, Lesina, Messina, Trieste (v. Graff), Sebastopol (Uljanin), Black Sea (Czerniavsky), Madeira (Langerhans, MSS.).

23. MACRORHYNCHUS CROCEUS (Fabricius, 9).

Length 1·5 mm. *Body* reddish, swollen posteriorly, pointed in front, where it is nearly white. *Proboscis* very powerfully developed. Of the parts composing the genital apparatus, the most diagnostic is the copulatory organ. At its proximal cylindrical end it receives the secretions of the vesicula seminalis and granule-reservoir, which are contained in an elongate sac strengthened by spirally-arranged muscles. The distal portion is spirally twisted, and consists of two canals, each containing a part of the continuation of the proximal single cavity. The upper edge of this spiral is "toothed." The egg-capsule is stalked. Several points, such as the relations of the germ- and yolk-glands and the presence of a spermotheca, are not yet satisfactorily determined.

HABITAT.—Among Fucus and Laminaria below the surface of the sea. Millport (v. Graff); Plymouth (F. W. G.).

DISTRIBUTION.—Apparently abundant in the northern seas. West coast of Greenland (Levensen) and of Norway (Jensen), Faroe Islands (Schmidt), Denmark (Oersted), Wimmereux (Hallez).

Among the Macrorhynchus collected at Plymouth were two species apparently new. Since, however, my observations are incomplete, I will not further describe them than by saying that one species closely resembled *M. mamertinus*, v. Graff, in the form and position of its gonads. The pharynx was not so strongly developed.

ii. VENENOSI.

24. MACRORHYNCHUS HELIGOLANDICUS, Metschnikoff (37).

Length ·5—1·5 mm. *Body* rounded at both ends, cylindrical, white, sometimes with brown spots. The *proboscis* is typical but small. The bilobed *brain* bears lenticulate *eyes*. The *pharynx* is rather small, placed as far in front of the centre of the body as the genital pore is behind it. The reproductive organs were first described by Jensen. The great variability of certain (especially the chitinous) parts, their complexity, and the presence or absence of certain accessory organs (spermotheca, &c.) according to the particular stage of development, render this perhaps the most difficult of all Turbellaria to elucidate. Personally I have found young specimens (·5—1 mm. in length) fairly intelligible. In these the *yolk-glands* form finger-shaped masses extending from the base of the pharynx to the genital pore. In the adult they become reticular and very bulky. The elongate, narrow germaria consist of a single row of ova for the greater part of their length. Behind their point of union is the large spermotheca. All these organs, the yolk-germ-glands and spermotheca, open into a single female genital canal. This canal is chitinised internally, and leads to the *genital atrium*. For an account of the male gonads with figures see v. Graff, 'Monographie,' pp. 330–1, pl. ix ; and Jensen, 'Turbellaria Norvegiæ.' pl. iv. The most important fact is that in addition to a chitinous sheath for the "granule-secretion," there is a common one for both this and the terminal vas deferens (see v. Graff, 'Monogr.,' p. 166, woodcut, fig. 9, *G.*). The *poison-organ* consists of a hollow chitinous *stylet* enclosed at its proximal end by a muscular sheath containing the poison-glands. A strong retractor muscle passes from the blind end of this muscular sheath, and is inserted on the upper end of the granule-reservoir.

HABITAT.—At the commencement of the Laminarian zone, Millport (v. Graff) ; Plymouth, Port Erin, Isle of Man (F. W. G.). Young specimens abounded at the last locality in October, 1892.

DISTRIBUTION.—West Greenland (Levinsen), White Sea (Mereschkowsky, 48), Bergen (Jensen), Wimmereux (Hallez).

A most remarkable character of the Proboscidæ, as a family, is the discontinuity of the gut caused by the development of the various genital organs, and in no form is this more conspicuous than in *Macrorhyncus heligolandicus*. In young specimens the gut is a closed sac surrounded by the body-cavity. As the gonads develop, becoming more and more bulky, the gut gets squeezed into any unoccupied spaces. Thus the gut-cells become scattered, and accumulate chiefly along the mid-dorsal surface. This fact accounts for the absence of a definite intestine in adult specimens.

Genus 14.—GYRATOR, Ehrbg., 1831.

Acrorhynchinæ with two genital pores, of which the female is the anterior.
  Germarium single ; testes elongate. Vesicula seminalis and
  granule reservoir separate, the latter with a special chitinous duct
  (v. Graff, 'Monogr.,' p. 331).

25. GYRATOR HERMAPHRODITUS, Ehrbg. (10).

    1875.   PROSTOMA LINEARE, McIntosh (45).
    1879.   „   „   Hallez (50).

Details of McIntosh's specimens are not given. I append the
following observations which I have made on specimens taken in the
neighbourhood of Manchester.

Length 1 mm. Body in the highest degree contractile, colourless,
cylindrical, tapering anteriorly. *Proboscis* very mobile. *Brain, eyes,*
and *pharynx* as in *Macrorhynchus*.

The genital organs are distinguished by the presence of a *poison-
dart* appended to the copulatory organ. Hallez has given a full
account of this with figures (see 43, pls. xx—xxii). The use of this
stylet as an offensive weapon has been seen by Schmidt ('Denkschr.
math.-nat. Klasse,' Wien, 1857) and Hallez ('Arch. Zool. Expt.,' 1873).
The animal bends the hinder end of its body towards the ventral
surface when close to its prey (small Entomostraca), which it stabs
repeatedly with its poison-dart.

HABITAT.—In sea water this form is only known from St. Andrews
under stones (McIntosh) and Maderia (Langerhans). It is widely
spread over Europe in fresh water.

Subfamily HYPORHYNCHINÆ.

" *Proboscis small, behind the anterior end, its sheath opening on the
ventral surface. Muscular cone present. Numerous short muscular
fibres constitute retractors. Spermatheca with chitinous appendage.
Vesicula seminalis and granule-reservoir not distinct. Their contents, however, issue through special chitinous ducts* " (v. Graff,
'Monogr.,' p. 336).

Genus 15.—HYPORHYNCHUS, v. Graff.

26. HYPORHYNCHUS ARMATUS (Jensen).

Length 1—1·5 mm. *Body* elongate, cylindrical, truncate at both
ends, white. Hinder end provided with strong *adhesive papillæ*. The
way in which these papillæ are used reminds one forcibly of a *Monotus*
(see p. 227). The anterior end, beset with long *flagella*, is moved
actively from side to side as it advances. Short *rhabdites* are present

  O

over the surface, modified on each side of the body behind the
middle into long vermiform bodies, in which Jensen perceived a
central thread. The opening of the *proboscis-sheath* is ventral, and
close to the anterior end. The *proboscis* itself is feebly muscular.
The *mouth* is a transverse slit, surrounded by an arcuate transverse
row of six adhesive papillae. The *eyes*, two on each side, lie over the
brain behind the proboscis. The *genital pore* is ventral, a short
distance from the hinder end. The *vasa deferentia* are given off from
the two rounded lateral *testes*. They unite along with the accessory
secretion at the base of the spirally coiled *ejaculatory duct*. This
consists usually of two coils and a terminal straight portion. The
spermotheca is armed anteriorly with chitinous spines; posteriorly,
according to Jensen, it communicates by a long narrow duct with the
genital atrium. (For figures see Jensen's "Turbellaria Norvegiæ" [49],
pl. iii, figs. 14—22).

HABITAT.—Among Zostera, Plymouth Sound; tide-pools, Port Erin,
Isle of Man (F. W. G.).

DISTRIBUTION.—Bergen (Jensen).

27. HYPORHYNCHUS PENICILLATUS (Schmidt, 32).

A young immature specimen that I refer rather doubtfully to this
species measured ·6 mm. in length. *Body* of a bright yellow colour,
the pigment being deposited in fine granules at the base of the
epidermal cells. *Rhabdites* small, occurring in numbers over the
surface. The aperture of the proboscis-sheath is triangular, ventral,
close to the anterior end. The *eyes* were fairly large, and provided
with lenses. The genital organs were not developed, hence the doubt
attaching to this example, which, however, in all remaining characters
agrees with *H. penicillatus* as described by v. Graff.

HABITAT.—Among Zostera, Cawsand Bay, Plymouth (F. W. G.).

DISTRIBUTION.—Lesina (Schmidt, 32), Messina, and Naples (v. Graff).

Family VORTICIDÆ.

*Rhabdocœla with a common genital pore. Germaria and vitellaria
united or distinct. Accessory female organs present. Uterus simple.*

Testes paired, compact. Mouth ventral, usually anterior. Pharynx dolioform.* Copulatory organ of various shapes (v. Graff, 'Monogr.,' p. 312).

Subfamily EUVORTICINÆ.

Pharynx and brain well developed. Germaria small. Body cavity extensive. Parenchyma small in amount. Free-living.

Genus 16.—PROVORTEX, v. Graff.

Euvorticinæ with two germaria and two distinct, elongate, unbranched vitellaria. Testes rounded. Pharynx dolioform. Mouth in the anterior third. Vesicula seminalis enclosed by the penis. Copulatory organ traversed by the spermatozoa (v. Graff. ibid., p. 344).

28. PROVORTEX BALTICUS (Schultze, 27).

Length ·6—1 mm. *Body* cylindrical, truncate in front, the angles produced into blunt processes, widening towards the middle and tapering behind to a long "tail." *Colour* brown, due to scattered reticular pigment. *Epidermis* containing *flagella* interspersed between the *cilia. Pharynx* provided with a distinct seam, into which the pharyngeal retractor muscles are inserted. From the extended observations of von Graff and Jensen it appears that this species is divisible into macro- and micro-pharyngeal varieties. Eyes paired, reniform. The *genital aperture* is ventral, at the base of the "tail." The *testes* lie far forward at the sides of the pharynx, and lead into paired vasa deferentia. These unite along with the "granule secretion" in the base of the copulatory organ. This organ is slightly variable in shape, and consists of a wide, cylindrical, tubular basal portion, opening through a narrow transverse slit, one edge of which is continued parallel to the axis of the cylinder, and then bends sharply at right angles. *Germ* and *yolk-glands* lateral; near their point of union is a curved *spermotheca*.

HABITAT.—This very active tiny animal occurred abundantly among Laminaria and also in brackish water at Millport (v. Gran). Abundant

* i. e. barrel-shaped. The term is, however, used in a technical sense, including certain structural peculiarities (see v. Graff, 'Monographie,' pp. 83—4).

in tide-pools, Port Erin, Isle of Man ; less commonly at Plymouth (**F. W. G.**).

DISTRIBUTION.—West coast of Greenland (Levinsen), West Norway (Jensen), Copenhagen (Fabricius, 9).

29. PROVORTEX AFFINIS (Jensen).

Length ·6 mm. *Body* stouter than in *Pr. balticus,* tapering posteriorly from the anterior third. *Pharynx* not so moveable as in the previous species. It is, however, in the form of the *copulatory organ* that these species are most easily and certainly distinguished. This is elongate, funnel-shaped, the terminal part of the duct bending at an obtuse angle with the proximal portion. Opposite this angle is a triangular plate projecting outwards from the surface of the duct.

HABITAT.—Along with *Pr. balticus,* Millport (v. Graff); Plymouth, along with *Monoporus rubropunctatus* (F. W. G.).

DISTRIBUTION.—Copenhagen (Fabricius), Bergen (Jensen).

30. PROVORTEX RUBROBACILLUS, n. sp. Pl. X, fig. 8 ; Pl. XI, fig. 12.

Length ·75 mm. *Body* cylindrical, broadly rounded in front, tapering slightly posteriorly. *Colour* mottled brown to the naked eye. The effect is due to numerous rods of doubtful nature in the interior of the gut-cells. They were present in all individuals examined. *Pharynx* without a distinct "seam." The free margin is crenulate. *Intestine* extensive. The *gut-cells* contains 3—8 rods of reddish colour. Whether they are zooxanthellæ (as described by v. Graff* in *Enterostoma zooxanthellæ*), or food-remains, is a moot point. Each eye possesses three lenses. The genital aperture is ventral, a short distance in front of the hinder end. A pair of *testes* lie at the sides of the pharynx ; they lead by wide *vasa deferentia* into a *vesicula seminalis.* The *penis* contains proximally the separate granule-secretion and *spermatozoa* separated ; distally it is converted into a chitinous copulatory organ, enclosing an inner muscular layer. The whole is bent into an S-shaped curve. As in *Pr. balticus,* one portion of the terminal

* 'Zool. Anzeiger,' 1886, p. 338.

margin is bent upon itself, and produced into an extremely fine, needle-like spine. *Germaria* and *vitellaria* as in *Pr. balticus.* A *spermotheca* is present near the genital pore.

HABITAT.—Dredged off the "New Grounds," Plymouth Sound (F. W. G.).

C. *ALLŒOCŒLA.*

The following definitions are v. Graff's (53), as amended by Böhmig (57, pp. 464–5):

Alimentary canal and parenchyma generally sharply separated; a body-cavity absent in the adult. Nervous and excretory systems present. Testes follicular. Germ- and yolk-glands separate or united, paired; the latter irregularly lobed, rarely branched. Gonads contained in parenchymatous cavities without a membrana propria. Penis formed by folds of the genital atrium. No conspicuous chitinous copulatory organ.

Family PLAGIOSTOMIDÆ.

Allœocœla with pharynx variabilis (except Plag. bimaculatum, where it is a pharynx plicatus), the size and position of which is subject to variation. Genital pore single or double; sometimes combined with the mouth. An otolith absent.

Sub-family PLAGIOSTOMINÆ.

Plagiostomidæ with a ventral and posterior genital aperture. Mouth anterior. Germaria and vitellaria distinct.

Genus 17.—PLAGIOSTOMA, O. Schmidt (28), 1852.
Without tentacles.

31. PLAGIOSTOMA DIOICUM (Metschnikoff, 37). Pl. XI, fig. 11.

Length ·6—·7 mm. *Body* cylindrical, tapering very slightly in front of the posterior third. *Colour* yellow-brown, lighter anteriorly and at the sides; eye pigment reddish brown. The *epidermis* contains a few small rods and numerous highly refractive vesicles. Böhmig (57, p. 408) has seen these in sections; they are possibly excretory.

Flagella are present anteriorly and posteriorly. *Mouth* anterior, subterminal. The *pharynx* is ellipsoidal, and lies altogether in front of the brain. The intestine is extensive, and corresponds generally to the form of the body. The *brain* is reniform, the slight concavity being directed anteriorly. A pair of nerves from the anterior and also from the hinder angles are conspicuous; three other pairs occur (Böhmig, 57, p. 410). Two *eyes* are present, provided with lenses. The *genital aperture* is ventral, and placed a short distance from the hinder end. The *testes* are scattered in the parenchyma. A muscular *vesicula seminalis* is placed close to the *genital pore*. The female organs are of considerable interest, as no one has yet found any trace of the *yolk-glands*. Since however, it is known that these organs develop late, it is possible that specimens presenting them in a mature condition have not yet been seen. If, on the other hand, yolk-glands are really not present at any stage, their absence would constitute a feature in which this species resembles the genus *Acmostoma*. The *ovaries* consist of a lateral row of clear spherical cells extending from the brain to the hinder end, and surrounded by refractive granules. I have not noticed the accumulation of ova behind the brain to which Böhmig refers (loc. cit., p. 316).

HABITAT.—Among littoral weeds, Plymouth (F. W. G.).

DISTRIBUTION.—Heligoland (Metschnikoff, 37), Trieste (Böhmig, 57). The specimens that I have seen appear to bear a close resemblance to *Acmostoma Sarsii*, Jensen. The form, colour, eyes, relations of the pharynx, character of the ova, testes, position of the vesicula seminalis, and apparent absence of vitellaria are almost identical in the two species. The presence of a narrow "creeping sole" in *Acmostoma* is, however, a distinguishing feature.

32. PLAGIOSTOMA SULPHUREUM, v. Graff. Pl. XII, fig. 20.

Length 2 mm. *Body* very elongate, cylindrical, parallel-sided for the greater portion of its length, somewhat conical in front, tapering posteriorly. *Colour* to the naked eye orange, the extreme anterior end paler; two large black eyes are conspicuous. *Movements* active, the front end being moved about as a flexible and highly sensitive "lip." The *tail* is provided with strong adhesive cells, by which the animal is securely fixed at will. The *epidermis* contains numerous *rhabdites* of a bright yellow colour, to which the tint of the animal is due. *Mouth* below, *pharynx* behind the brain. The pharynx is very small, the musculature being slightly developed. Numerous glands surround it. The *intestine* occupies the central part of the body, and is enclosed anteriorly by spherical glands. The *genital aperture* lies a short distance from the hinder end on the ventral surface. The small *testes* are placed behind the centre of the body in the middle line. The *spermatozoa* have a very characteristic form. They are divisible into a broad head, and a narrow pointed tail. A dark transverse band separates the head end off as a pointed lid. Down the centre runs a spiral thread. The *ova* develop from a median cellular mass which, according to Böhmig (57, p. 365), lies close to the brain. In compression preparations it is driven posteriorly (Pl. XII, fig. 30). The developing ova lie at the sides of the gut. The *yolk-glands* are paired, large, lobed organs, more or less enclosing the intestine and uniting behind the pharynx.

HABITAT.—In tide-pools, among corallines, Port Erin, Isle of Man (F. W. G.).

DISTRIBUTION.—Trieste (v. Graff and Böhmig).

33. PLAGIOSTOMA ELONGATUM, n. sp.

Length 2 mm. *Body* cylindrical, stout, elongate, rounded in front, tapering rather suddenly posteriorly. *Colour* opaque white. *Epidermis* provided with cilia, which are longer anteriorly than elsewhere, and between which stiff flagella occur. Narrow oblong rhabdites are thickly scattered over the surface. They are homogenous and highly refractive. The musculature is strongly developed, and this fact, combined with the opacity of the other organs, renders it a matter of difficulty to examine this species by compression. The *mouth* is subterminal. *Pharynx* large, barrel-shaped when at rest, situated behind the brain. It is very muscular, and can extend under and in front of the brain towards the mouth. The *intestine* is large, corresponding generally to the form of the body, slightly hollowed in front to receive the base of the pharynx. Two larger irregular black eyes unite by strands of pigment across the brain. The *genital pore* is close to the hinder end. The male organs were not thoroughly developed in either of my two specimens, and the form of the mature *spermatozoa* remains unknown. The *germaria* lie at the sides of the posterior third of the body. The *yolk-glands* form narrow-lobed masses at the sides of the gut, extending as far forward as the base of the pharynx.

HABITAT.—From coarse sand at the bases of *Corallina officinalis*, Plymouth (F. W. G.).

34. PLAGIOSTOMA PSEUDOMACULATUM, n. sp.

Length 2 mm. *Body* elongate, pointed behind, the anterior end not distinctly separated off from the rest of the body; hence it may be distinguished from *Pl. maculatum*, which this species closely resembles in form and in many points of structure. *Colour* white, a violet patch of reticular pigment between the eyes. *Mouth* lies beneath the brain. *Pharynx* very muscular. The *genital aperture* is ventral, at the base of the tail. The *germaria* lie right and left behind the pharynx; behind these again the *testes*. The *vasa deferentia* are distinctly swollen before entering the cylindrical muscular *penis*.

HABITAT.—This species, so far as my experience goes, belongs to a deep-channel fauna of Plymouth, characterised more especially by the presence of various *Monotida*, to be presently described. I have found it always associated with *Polydora ciliata* (Polychaeta), which forms mud-tubes in hundreds on the muddy bottom of the Hamoaze.

This species differs from *Plagiostoma maculatum* in the absence of the red intestine and lateral head-grooves which characterise the latter.

35. PLAGIOSTOMA SAGITTA (Uljanin, 41).

Length 1 mm. *Body* elongate, conical in front, tapering gradually backwards from the middle. The tail not so long as in the preceding species. *Colour* opaque white with a slight yellowish tinge, due to the contents of the gut. Rhabdites, grouped in clumps, are present over the surface. *Pharynx* behind the *brain*, which is transversely elongate, deeply incised anteriorly. Two pairs of eyes are present, placed over the brain; the hinder pair is the larger, and is markedly reniform. Genital aperture at the base of the tail. The *vesicula seminalis contains spermatozoa*, which have a central rib, bearing broad triangular lateral membranes, exactly similar to the spermatozoa of *Plagiostoma maculatum*. Contrary to Uljanin (41), I find two *vesiculæ present*. They lie at the middle of the sides of the body.

HABITAT.—In 5 fms., among harbour débris, Plymouth (F. W. G.)

DISTRIBUTION.—Sebastopol (Uljanin).

36. PLAGIOSTOMA CAUDATUM, Levinsen (51).

Length 1·5 mm. *Body* when swimming cylindrical, tapering from the middle posteriorly, conical in front. *Colour* yellow, due to epidermal granules. *Rhabdites* few. *Pharynx* behind the brain. *Eyes* large, rhomboidal; between them a small mass of reticular pigment, with which, according to Levinsen, the pigment-cups of the eyes are sometimes connected. The *genital aperture* lies a short distance in front of the hinder end. The *seminal vesicle* leads by a narrow duct into the cylindrical muscular *penis*, surrounded by a double sheath, as in *Pl. reticulatum*.

HABITAT.—In 5¼ fms., Plymouth Sound (F. W. G.).

DISTRIBUTION.—Egedesminde, west coast of Greenland (Levinsen).

37. PLAGIOSTOMA VITTATUM (Frey v. Leuckart. 23).

Length 1—2 mm. The latter size is that of individuals which have just laid their cocoons. v. Graff ('Monogr.,' p. 391) states that a length of 3 mm. is reached by specimens living on Laminaria below the surface. Jensen also (49, p. 58) finds larger specimens in such localities than at the surface. Tow-nettings taken near shore at Plymouth yielded small examples. *Body* convex above, broadly rounded in front, tapering to a finely pointed "tail" behind. It is well figured by van Beneden (33), pl. v, figs. 1 and 2. *Colour* more variable than in any other species of *Plagiostoma*. The typical coloration is three transverse bands of violet reticular pigment on a white ground; one central, one across the head, and the third across the tail. v. Graff, in his Monograph, pl. xviii, fig. 6, has figured eight different varieties of the arrangement of these bands, which he observed in a single gathering among Ulva at Millport ; and this does not exhaust the possible cases. Small specimens (·5—1 mm.) with a yellow ground-colour (due apparently to the contents of the gut-cells) are not uncommon. Since the "key" for the determination of species of this genus in v. Graff's Monograph is largely dependent on the arrangement of the pigment, these varieties are at first very troublesome. More-over *Vorticeros auriculatum* (which frequently occurs among *Plagiostoma vittatum*) with retracted tentacles, can be in no way anatomically distinguished from a common variety of *Plag. vittatum*, in which the pigment is present over the greater part of the dorsal surface. With regard to the internal anatomy I can confirm the accounts of v. Graff and Jensen (49). Van Beneden (33), the first to discover the stalked yellow-brown egg-capsules, found them attached to the abdominal feet of the lobster. At Plymouth they were abundant in September on the sides of a vessel containing several individuals.

HABITAT.—Apparently more abundant on our northern than southern coasts. Millport, abundant (v. Graff) ; Plymouth ; Port Erin, Isle of Man (F. W. G.).

DISTRIBUTION.—Faroe Islands (Schmidt), Bergen (Jensen), Heligo-land (Leuckart, 23), Walcheren, on coast of Belgium (Slabber),[*] Ostende (van Beneden), Wimmereux (Hallez).

38. PLAGIOSTOMA KORENI, Jensen.

Length 1·4 mm. *Body* similar in shape to *Plag. vittatum*, but smaller, and rather narrower in front. The first half of the body is white, and behind this a broad transverse brown band (due to reticular pigment) over the dorsal surface and the sides. Behind this again are scattered brown spots. The *brain*, which bears two red *eyes*, is placed in front of the spherical *pharynx* and *mouth*. The remaining anatomy has been investigated by Jensen (49), whose figures (pl. v, figs. 1—8) are highly characteristic.

HABITAT.—On the inner side of the Breakwater, and elsewhere among algæ between tide-marks, Plymouth Sound (F. W. G.); a specimen at Millport (v. Graff).

[*] 'Physikalische Belustigungen,' Nürnberg, 1775, pp. 31 and 36.

Distribution.—Bergen (Jensen).

30. ?Plagiostoma siphonophorum (O. Schmidt, 28).

Length ·9 mm. *Body* elongate, truncate in front with rounded angles, tapering posteriorly. Along the mid-dorsal line is a narrow band of reticular black pigment, which expands laterally behind the eyes and extends between and beyond them anteriorly. Böhmig has shown (57, pp. 208–9) that this pigment is present in the gut-cells and not in the parenchyma. The *mouth* lies beneath the brain. The *pharynx*, when extended through the mouth, is narrow, cylindrical, expanded slightly at the distal end. My observations on the genital organs are incomplete. The *penis* is pyriform, armed at its base with small chitinous spines.

Habitat.—In 15 to 18 fms., Plymouth Sound (F. W. G.).

Distribution.—Lesina (Schmidt), Trieste (v. Graff, Böhmig).

As ripe sperm was not seen, this species remains doubtful. But other characters agree with *Pl. siphonophorum*, so I refer my specimen to this form.

40. Plagiostoma Girardi (Schmidt, 32).

Length 1·75–2 mm. Southern examples range up to 3 mm. *Body* slightly depressed, rounded in front, tapering behind, broadest in the middle. v. Graff's figure ('Monogr.,' pl. xviii, fig. 12) does not represent the form well, whereas, as Böhmig has remarked, the figure of *Pl. ochroleucum* would do so. *Colour* white. *Movements* sluggish. *Rhabdites* abundant over the surface. *Pharynx* small, opening through the mouth behind the brain. *Eyes* two, reniform, with lenses. According to Böhmig (57, p. 355), the "*ciliated furrow*" forms a transverse groove in front of the mouth, and does not correspond to a slight constriction which occurs behind the eyes. For the genital organs the accounts of v. Graff and Böhmig may be consulted. The *spermatozoa* consist of a midrib, bearing broad triangular hyaline membrane, and terminating in a short anterior and a longer posterior flagellum.

Habitat.—Not uncommon 6–15 fms., Plymouth Sound (F. W. G.)

DISTRIBUTION.—Naples (v. Graff), where it is the most abundant Rhabdocœle ; Trieste (v. Graff and Böhmig) ; Messina (v. Graff).

41. PLAGIOSTOMA OCHROLEUCUM, v. Graff.

Length 5·5 mm. *Colour* whitish yellow. *Mouth* sub-terminal. *Pharynx* very small, beneath and in front of the brain. Remaining anatomy similar to *Pl. Girardi*.

HABITAT.—1½ fms, among Laminaria, Millport (v. Graff).

Genus 18.—VORTICEROS, O. Schmidt (28), 1852.

Plagiostomina with two tentacles at the anterior end.

42. VORTICEROS AURICULATUM (O. F. Müller, 4).

Length 1·5 mm. *Body* produced anteriorly into a pair of long tentacles. Behind the eyes it is slightly constricted, and then expands towards the middle, tapering behind to a fine point. The *tentacles* are not often seen fully extended. At the slightest alarm they can be completely withdrawn, and the animal may continue to swim about in this condition. The *colour* is variable, but usually consists of a broad band of dark carmine reticular pigment on the upper surface, leaving the side margins free, and continued on to the tentacles. The anatomy differs in no important particulars from that of the genus *Plagiostoma*.

HABITAT.—Among Ulva and other littoral weeds, Plymouth ; Port Erin, Isle of Man (F. W. G.) ; Millport (v. Graff).

DISTRIBUTION.—Naples, Trieste, Messina (v. Graff), Wimmereux (Hallez), Norwegian coast (Müller).

43. VORTICEROS LUTEUM, v. Graff (53).

1852. VORTICEROS PULCHELLUM, O. *Schmidt* (28).
1879.   ,,   ,,   var. LUTEUM, *Hallez* (50).

This species, established by v. Graff for the reception of a large specimen (8 mm. long), is distinguished from the preceding by its stouter appearance and uniform yellow colour. On two occasions at Plymouth an example measuring 2·5 mm. in length was taken. One was found among *Bugula turbinata* from 7 fms., the other among littoral weeds at a low spring-tide.

DISTRIBUTION.—Wimmereux (Hallez), Naples (v. Graff).

Sub-Family ALLOSTOMINÆ.

Plagiostomidæ with one ventral and posterior genital aperture. Two germaria and two distinct vitellaria. Pharynx placed in the hinder half of the body, its mouth directed posteriorly.

This subfamily, as constituted by v. Graff, includes the genera *Enterostoma* and *Allostoma*. Concerning it our knowledge is in a most unsatisfactory condition. With regard to the former genus we do not possess a good description of any species, although these are abundant in northern and southern seas. Consequently the definition given above must be considered provisional. Already *Enterostoma striatum,* v. Graff, has been investigated, with the result that Böhmig places it under the sub-family *Cylindrostominæ* (loc. cit., p. 469), and similar change in species hitherto considered as belonging to the *Allostominæ* may be expected to result from detailed investigations.

Genus 19.—ENTEROSTOMA, Claparède (35).

Allostominæ with uniformly ciliated body and without a circular "ciliated groove" on the head.

44. ENTEROSTOMA AUSTRIACUM, v. Graff. Pl. X, fig. 7.

Length ·75 mm. *Body* rounded anteriorly, tapering to a blunt "tail" behind. *Colour* usually yellow, with a black spot (due to the intestine) in the centre; occasionally the general colour is white, the gut being yellow. The colour of the surface is due to the presence of yellow granules in the epidermis; that of the gut to its contents. The yellow epidermal granules are massed together in small heaps, which are not so conspicuous or large as figured by v. Graff (pl. xix, fig. 9, Monograph). Exceedingly small, slender *rhabdites* are present in groups in the epidermis. The *pharynx* is cylindrical, very muscular, and is inserted at the hinder edge of the extensive gut, which, following the outline of the body, reaches as far forward as the brain. The *eyes* are arranged in two pairs, an anterior and a posterior pair. The former are slightly the smaller, and their lenses are directed outwards and backwards; the lenses of the posterior pair face forwards and outwards. The genital aperture is a short distance from the hinder extremity. Round the brain are the developing *testes*. The pyriform muscular *penis* lies behind the pharynx.

HABITAT.—In 4—18 fms., Plymouth and Port Erin (F. W. G.)

DISTRIBUTION.—Trieste (v. Graff).

45. ENTEROSTOMA FINGALIANUM Claparède (35).

Length 1 mm. *Body* elongate, cylindrical, rounded at both ends. *Colour* white ; the short, almost central gut reddish. The epidermis contains small fusiform rhabdites, figured by Hallez (50, pl. ii, fig. 25). *Mouth* at the commencement of the posterior third. *Pharynx* cylindrical at the hinder margin of the intestine. *Brain* bilobed. *Eyes* fairly large, arranged as in *Ent. austriacum*. *Genital pore* between the mouth and the hinder end. *Testes* numerous, surrounding the brain. Vasa deferentia exhibit dilatations along their course. *Penis* pyriform, containing masses of accessory secretion in its basal portion ; provided at its tip with small papillæ. Claparède's original description and figures are not sufficiently distinctive to enable us to determine whether the *germ-* and *yolk-glands* are distinct or not. In my specimen the *yolk-glands* form lateral masses uniting behind the brain, and again posteriorly behind the genital aperture. The germ-glands lie at the sides of the pharynx, and appear to be separate from the yolk-glands. As, however, only a single specimen was available, further observations are greatly needed on this and other points, such as the presence of a ciliated "head-furrow" (which Böhmig has demonstrated in other forms by the use of methylene-blue in cases where superficial observation had previously failed) and the possible connection of genital and oral apertures.

HABITAT.—Skye (Claparède) ; among Balanus, 10 fms., Plymouth Sound (F. W. G.).

DISTRIBUTION.—Wimmereux (Hallez).

46. ENTEROSTOMA CÆCUM, v. Graff.

Length 1·7 mm. *Body* gradually tapering forwards from the hinder end. Beneath the *epidermis* are yellowish-green granules, especially abundant at the sides. The *pharynx* is cylindrical, muscular, placed far posteriorly. The *spermatozoa* consist of a central rib bearing lateral membranes, and produced into a fine flagellum in front and behind. *Eyes* absent. For further details and figures see v. Graff, 'Monogr.,' p. 101, and pl. xix, figs. 15—17.

HABITAT.—A specimen at Millport in a tide-pool (v. Graff).

v. Graff described this form as the only blind Plagiostomid. I have, however, found a probably new species of Plagiostoma in which the eyes were wanting.

Genus 20.—ALLOSTOMA, van Beneden (33).

Allostominæ in which the " circular furrow " at the level of the brain is provided with long cilia.

47. ALLOSTOMA PALLIDUM, van Beneden (33).

Length 2 mm. *Body* cylindrical, tapering slightly towards each extremity. The anterior sixth is sharply separated from the rest of the body by a transverse marking, the nature of which is not quite clear. It is probably due to the ciliated "circular furrow." *Colour* yellowish white. The *epidermis* contains numerous pseudo-rhabdites. Considering rhabdites as a condensed glandular secretion, *pseudo-rhabdites* are intermediate between the amorphous secretion and rhabdites. The *mouth* is posterior ; the *pharynx* short, leading into an extensive gut. v. Graff has described the genital organs fully. *Testes* surround the brain. *Vasa deferentia* convey the sperm in balls to the base of a pyriform *penis*, in front of which lie the ovaries. The *oviducts* unite and open through the subterminal genital pore. *Yolk-glands* lateral, lobed. (See v. Graff, 'Monogr.,' pl. xxix, figs. 12—14.)

HABITAT.—Millport (v. Graff).

DISTRIBUTION.—Ostende (van Beneden, v. Graff).

Van Beneden (33) has described the oval egg-capsules, which are very small, and extruded one at a time. The young when hatched are without a definitive pharynx, gut, eyes, or brain. They become sexually mature in three weeks.

Subfamily CYLINDROSTOMINÆ.

v. Graff's definition of this sub-family ('Monogr., p. 409) has been

materially altered owing to Böhmig's researches.   It now reads thus
(Böhmig, 57, p. 469):

*Plagiostomida with a ciliated " circular groove."   The oral and genital
apertures combined.   A germ-yolk-gland present.   Spermotheca present,
connected with the ovigerous cell-mass (" Keimlager").*

Genus 21.—CYLINDROSTOMA, Oersted (21).

The limits of this genus are not yet satisfactorily defined.   v. Graff
divided it into prosoporous and opisthoporous forms, according as the
mouth was anterior or posterior.   The latter have been excluded by
Böhmig in his definition of the genus.   v. Graff's original extension
of the genus is, however, here adopted, pending a thorough examination
of the *Opistoporia*.

48. CYLINDROSTOMA QUADRIOCULATUM (Leuckart, 23).

Length ·5—·8 mm.   *Body* colourless, somewhat depressed, rounded
in front, tapering posteriorly to a long "tail" beset with adhesive
cells.   *Mucus-rods* ("Schleim-stäbschen"), of an irregular granular
character, occur in the epidermis.   *Flagella* are interspersed among
the cilia in front and behind.   At the level of the brain a pair of well-
marked ciliated furrows are present.   *Mouth* ventral, in front of the
brain.   *Pharynx* elongate cylindrical, extending from the brain to the
centre of the body ; its anterior margin is crenulate, and provided
with stout flagella.   *Brain* almost cubical.   The *genital aperture* is
combined with the mouth.   This remarkable discovery, made by
Böhmig, corrects former mistakes due to misleading compression
preparations.   The *testes* form large follicular masses surrounding the
brain.   The different stages in the development of the *spermotozoa*
can be well observed.   Behind the *penis* are a pair of short, wide *vasa
deferentia*, which open into a highly glandular vesicula seminalis.
The *penis* is a muscular, cup-shaped organ receiving both spermatozoa
and granule-secretions from the vesicula, which it transmits through
the genital atrium (underneath the pharynx), and so to the exterior.
The *spermatozoa* are elongate, wider in front than behind.   The
central axis of the tale is markedly granular, and is continued forwards
as a spiral thread, wound three times round the surface of the " head."

The posterior end of the body is occupied by the large spermotheca. The *germ-yolk-gland* is composed of an anterior vitelline, and a posterior germinal portion.

HABITAT.—Kilmore, Skye (Claparède); abundant in tidepools, Millport (v. Graff); among *Ptilota*, *Ceramium*, and other algae. Plymouth (F. W. G.).

DISTRIBUTION.—Faroe, west coast of Norway (Claparède and Jensen), Heligoland (Leuckart), Ostend (van Beneden), Sebastopol (Uljanin).

49. CYLINDROSTOMA FELINUM (Hallez, 50). Pl. X, fig. 4.

Length 1 mm. *Body* oval, broadly rounded in front, tapering behind, of a bright yellow colour. Opposite the level of the brain, right and left, are a pair of lateral grooves, bordered by long cilia. The *epidermis* contains masses of granular yellow pigment and rhomboidal rhabdites. *Mouth* ventral, behind the brain. The *pharynx* is cylindrical, with a crenulate anterior margin. The genital organs resemble those of *Cyl. quadrioculatum* very closely; a *spermotheca*, however, is absent.

HABITAT —Among fine red seaweeds, Plymouth (F. W. G.).

DISTRIBUTION.—Wimmereux (Hallez).

This species exhibits very great similarity to *Cyl. Klostermanni*, Jensen, especially in form, colour, and general anatomy. The chief points of distinction are the absence of calcareous bodies in the epidermis, and of a spermotheca.

50. CYLINDROSTOMA ELONGATUM, Levinsen. Pl. XII, fig. 19.

Length ·6—·8 mm. *Body* very elongate, narrow, cylindrical, slightly tapering and rounded in front, pointed behind. An apparent groove was seen just above the level of the brain, separating off the portion of the conical head in front of it. *Colour* yellowish to the naked eye : a black spot (the intestine) lies in the centre. The epidermis bears specially long cilia at the extremities. Small *mucus-rods* are present in large numbers. Glands open at the anterior end in front of the brain. *Mouth* ventral, posterior. *Pharynx* attached to the hinder end of the gut, barrel-shaped, its free margin crenulate. The *intestine*

P

has a very small longitudinal extent, not greatly exceeding the transverse diameter of the body. It contains yellow-green and reddish-brown remains (chiefly diatoms). *Brain* cuboidal, the angles rounded off, without fissures. Four *eyes*, the posterior pair being distinctly the larger. Genital aperture almost terminal, just under the anterior end. *Testes* eight to nine in number, in front and at the sides of the brain. *Vasa deferentia* lead to the base of the posteriorly-directed, pyriform penis. Numerous glands open at this point, and their secretions are arranged in a radiate way. The *germ- yolk-glands* are bulky, and lie at the sides of the gut; they unite behind the brain. Behind and at the sides of the pharynx is the ovarian portion of the gland.

HABITAT. — Among tide-pools, Wembury Bay, near Plymouth (F. W. G.).

DISTRIBUTION.—Egedesminde, Greenland (Levinsen).

Genus 22.—MONOOPHORUM, Böhmig, 1891.

Cylindrostomina with united mouth and genital apertures. Pharynx directed backwards, the penis forwards. The spermotheca opens into the genital atrium. The germinal portions of both germ-yolk-glands are fused together in the middle line dorsally.

51. MONOOPHORUM STRIATUM (v. Graff).

Length 1 mm. *Body* cylindrical, rounded in front, pointed behind. *Colour* carmine to the naked eye. Under the microscope, however, it is seen that the reticular pigment is well developed, leaving the margins of the body and the outer sides of the eyes almost free. The surface of the body has a characteristic "streaked" appearance, caused by the grouping of the longitudinal muscles into bundles of four to six. In the intervals small rhabdites are plentiful. The *cilia* are very strongly developed. Böhmig has discovered that the oral and genital apertures unite. The *pharynx* is very contractile. The *spermatozoa* are collected in a pair of *vasa deferentia* and transferred to the globular base of the *penis*, the terminal part of which is narrow and cylindrical. For a more detailed account see Böhmig (57), pp. 435—447.

HABITAT.—Dredged in a few off the Duke Rock, Plymouth Sound (F. W.   ).

DISTRIBUTION.—Trieste, among Ulva (v. Graff, Böhmig).

Family MONOTIDÆ.

Alæocœla with two genital apertures. A spermotheca present. Two germaria   distinct vitellaria. Testes follicular, closely   between the pharynx and the brain. Pharynx   An otolith present. Elongate flat forms, with a narrow   broad posterior extremity furnished with "   "

Genus 23.—MONOTUS, Diesing.*

Monotidæ in which the female genital pore lies in front of the male.

52. MONOTUS LINEATUS (O.   )

    1773. FASCIOLA LINEATA, *O. F. Müller* (2).
    1853. PLANARIA FLUSTRÆ, *Dalyell* (29).
    1861. MONOCELIS LINEATA, *Claparède* (35).
    1861. MONOCELIS AGILIS, *Claparède* (35).
    1865. TYPHLOPLANA FLUSTRÆ, *Johnston* (38).
    1875. MONOCELIS RUTILANS, *McIntosh* (45).
    1882. MONOTUS LINEATUS, *v. Graff* (53).

This synonymy refers merely to the works of authors who have described British examples of this species.† For a fuller list see v. Graff (53), p. 418.

Length 2—2·5 mm. *Body* very elongate, appearing to the naked eye as a fine white thread. The hinder end assumes the form of a disc when the animal contracts. By means of adhesive papillæ present on the surface of this "Haftscheibe" it clings very tenaciously to the substratum. *Colour* variable, sometimes absent, more frequently present in the form of brown or grey reticular pigment. The *epidermis* of the anterior end is markedly thicker than elsewhere, and bears numbers of well-developed sensitive *flagella*. This part of the body is constantly employed during life in active movements in all directions.

* Diesing, K. M , "Revision d. Turbellaria, Rhabdocœlen," ' Sitzungb. d. Akad. Wien,' Bd. xlv, 1862, p. 211.

† A method adopted throughout this memoir.

Should it meet with an obstacle it retracts with amazing rapidity. The *rhabdites* are only feebly developed. Owing to the great contractility of the body the positions of the organs are difficult to define. Considering, however, the animal to be in a fully extended state, the mouth is a short distance behind the centre of the body. The *pharynx* is cylindrical, very muscular, its proximal end being almost central. The gut is extensive; when contracted it becomes distinctly sacculated. In the middle line anteriorly is the *otolith*, composed of a vesicle containing a central concretion bearing two double lateral ones. Immediately in front of this is the single transverse brown "eye," and behind it the brain. The *male genital pore* lies at the commencement of the adhesive "tail;" the female pore between this and the pharynx. The *testes* are numerous. The *vasa deferentia* run back to a muscular vesicula which opens into a papilla surrounded by accessory glands. This soft papilla is the *copulatory organ*. The single pair of *germaria* lie at the base of the pharynx. The *vitellaria* occupy the sides of the body.

HABITAT.—Hebrides (Claparède, 35); Firth of Forth, "on *Flustra hispida*" (Dalyell, 29); Millport (v. Graff); St. Andrews (McIntosh, 45); Port Erin, Isle of Man; Plymouth (F. W. G.).

DISTRIBUTION.—Very wide. West coast of Greenland (Levinsen), south and west coast of Norway (Claparède), Baltic (Müller, Schultze), North Sea (van Beneden), Madeira (Langerhans), Naples, Messina, Trieste (v. Graff), Black Sea (Uljanin, Czerniavsky).

This species is readily distinguished from *M. fuscus* by its unarmed penis.

53. MONOTUS FUSCUS (Oersted, 16).

Length 1·5—3 mm. *Form* similar to *M. lineatus. Colour* very variable, usually brown, but white, purple, and even dark blue varieties have been recorded by Jensen and v. Graff. Examples 1 mm. in length are usually white and colourless. Larger specimens (2 mm.) are frequently carmine, gradually becoming brown as they grow older. The meaning of this change[*] in the reticular pigment is not understood. Similar changes in some Opisthobranchiate Molluses are

* Already remarked by v. Graff (53), p. 422.

known (Aplysia, see Garstang, 'Journal Marine Biol. Assoc.,' N.S.,
vol. i. No. 4, p. 411) ; and in this case the change in colour appears to go
hand in hand with a change of surroundings. *Pharynx, brain, otolith,*
and *eye* as in *M. lineatus.* The *male pore* is placed further forward
than in the latter species. The *vasa deferentia* open into the neck of
the very muscular *vesicula seminalis.* The *copulatory organ* is a
hollow chitinous spine of variable shape, connected at its base by
muscles to the wall of the seminal vesicle. Numerous accessory
glands open at this level. The *spermatozoa* are whip-shaped, the
handle being stout, the lash a very fine thread. Opening to the
exterior through the female genital pore is the *spermotheca*, provided,
according to v. Graff and Jensen, with secondary lobes. The remaining
parts of the female genital apparatus do not materially differ from
those of *M. lineatus.*

HABITAT.—This species extends its range to the higher parts of the
littoral zone. In consequence it is liable to be exposed to the air for
some hours. Many of its devices for obtaining a moist position
during ebb-tide have been described by Hallez and v. Graff. Thus the
former observer collected *Balani*, the latter *Chitons* and *Patellæ* at
low tide. After placing these in sea water, *Monotus fuscus* crept out
of gills or thoracic limbs as the case might be. In the Isle of Man I
have found them nestling among the appendages of Balani. Millport
(v. Graff) ; Port Erin, and Plymouth (F. W. G.).

DISTRIBUTION.—Faroe Islands) Schmidt, 24), Bergen (Jensen, 49),
Drobeck and Denmark (Oersted, 16), Heligoland (v. Graff), Cuxhaven
in the Baltic (Schultze), Osteud (van Benedin [33] and v. Graff).
Wimmereux (Hallez, 50).

54. MONOTUS ALBUS, Levinsen (51).

Length 1·3 mm. (*i.e.* half that of Levinsen's specimen). *Body*
elongate, narrow, the hinder end not expanded into a disc, colourless ;
the contents of the gut reddish. Ocular pigment absent. *Pharynx*
posterior. A large *spermotheca* containing a refractive secretion opens
to the exterior in front of the penis, which is armed with a shoe-
shaped chitinous *copulatory organ*, bearing a couple of lateral teeth on
the free margins.

HABITAT.—One specimen in a tide-pool, Plymouth (F. W. G.).

DISTRIBUTION.—Jacobshavn, West Greenland (Levinsen).

Genus 24.—AUTOMOLOS, v. Graff (53).

Monotida in which the female pore lies behind the male.

55. AUTOMOLOS UNIPUNCTATUS (Oersted, 16).

1826.   PLANARIA UNIPUNCTATA, *Fabricius* (9).
1844.   MONOCELIS UNIPUNCTATA, *Oersted* (16).
1851.   „   „   *Schultze* (27).
1861.   „   SP. (? UNIPUNCTATA, Oe.), *Claparède* (35).
1875.   „   UNIPUNCTATA, *McIntosh* (45).
1878.   „   SPINOSA, *Jensen* (49).

Length 1—1·5 mm.   *Body* resembling *Monotus fuscus* in form, the hinder end, however, not expanded into an adhesive disc.   Usually colourless.   Ocular pigment is absent.   *Otolith* with a pair of simple accessory concretions.   The *mouth*, *pharynx*, and *intestine* resemble those of the preceding species.   My specimens, as might be concluded from their small size (Jensen's were 3 mm., Schultze's measured as much as 6·6 mm.), were immature, and in consequence the genital ducts were not fully developed.   According to the naturalists just named, the *penis* lies behind the male genital pore.   It consists of a *vesicula seminalis* which conveys both spermatozoa and accessory secretions into the coiled *ductus ejaculatorius*, the terminal portion of which when extended is finger-like, and provided with small spines of variable shapes on its exterior.   (See Schultze [27], pl. ii, and Jensen [49], pl. vi, fig. 9.)   The two oviducts unite in the anterior region of the body, and the common duct runs back to the female genital pore, which also receives the duct of a vesicle—apparently spermotheca and uterus combined, since Jensen found sperm and ova in it.

HABITAT.—Skye (Claparède); St. Andrews, under stones between tide-marks (McIntosh); among littoral algæ, Plymouth (F. W. G.).

DISTRIBUTION.—Bergen (Jensen), coast of Denmark (Fabricius, Oersted), Greifswald (Schultze), Madeira (Langerhans), Black Sea (Uljanin Czerniavsky).

56. AUTOMOLOS HORRIDUS, n. sp.   Pl. XII, fig. 21.

Length 1·5 mm.   *Body* somewhat flattened.   A slight constriction occurs at the level of the otolith, separating off an anterior conical portion.   Behind this " neck " the body gradually increases in width, to the posterior third of its length which ends in a sharply pointed "tail."   Pigment is absent, the gut alone giving a grey tinge to the otherwise white body.   *Flagella* are present at each extremity, and also occur at intervals for a short distance behind the anterior end.   Packets of *Rhabdites* occur in large numbers on the surface of the body. (fig. 21).   *Adhesive cells* are present, although feebly developed on the tail.   The musculature is strong, enabling the animal to execute very active movements, and to flex the sides of the body ventrally towards the middle line.   The *mouth* lies at the commencement of the posterior third.   The *pharynx*, inserted into the gut at the centre of the body, is cylindrical and very muscular.   The *intestine* lies chiefly in front of the pharynx.   Its cavity is produced into about twelve cæca on each side, placed fairly symmetrically.   The specimen under examination was starved, and in this condition the limit of the gut branches can be clearly defined.   *Brain* placed behind the otolith, oval, the long axis coinciding with that of the body.   An eye is absent.   *Testes* in this specimen not well developed.   They occur behind and at the sides of the brain.   *Vasa deferentia* open behind the pharynx into a *vesicula seminalis*, surrounded by accessory glands.   The *penis* is pyriform and muscular.   The single pair of *germaria* lie opposite the base of the pharynx.   Elongate *yolk-glands* are placed at the sides of the gut.   An accident prevented the determination of the oviducts.

HABITAT.—One specimen dredged in 12 fms., Plymouth Sound (F. W. G.).

57. ? AUTOMOLOS OPHIOCEPHALUS (O. Schmidt).   Pl. XI, fig. 18.

> 1861.   MONOCELIS OPHIOCEPHALA, *Schmidt* (34).
> 1882.   AUTOMOLOS OPHIOCEPHALUS. *v. Graff* (53).

Length 1·5 mm.   *Body* of a pink colour, very slender, elongate, the anterior end broader than the rest of the body, and separated off by a slight constriction.   The hinder end when freely moving tapers gradually to a point.   During contraction it becomes thickened and widened. *Flagella* appear to be absent. The rods are accumulated in packets chiefly at the two extremities. Individual rhabdites are longer at the anterior, shorter at the posterior end—exactly the reverse of the case in *A. hamatus*, Jensen. Strong adhesive cells occur on the " tail." The *pharynx* is placed about the commencement of the posterior third. When extended the free end expands, the base becoming constricted. The intestine, which contains pinkish granules, is marked sacculated. The pouches numbered about twenty on each side, and in compression-preparations appeared to be fairly definitely paired. Between successive gut-sacs were muscular dissepiments. Ocular pigment is absent. *Testes* occupy spaces in front and at the sides of the pharynx. The *vasa deferentia* lead to a *vesicula seminalis*, and this opens, along with the accessory glands, into the *penis*, a pyriform muscular organ, similar in position and form to that of *A. hamatus*. The pair of *germaria* lie at the base of the pharynx, the *yolk-glands* accompanying the gut-pouches and lying between them. The *oviducts* were not observed.

HABITAT.—Dredged in twenty fms., Plymouth Sound (F. W. G.).

DISTRIBUTION.—Corfu (O. Schmidt).

Schmidt's description of this species does not agree in all points with the diagnosis just given of my own specimen. The differences consist in the following details : the presence of ocular pigment in front of the otolith, and the relations of the pharynx and ovaries. The latter, in his specimen, occupied a position behind and not in front of, the pharynx. The extreme contractility of the pharynx itself, and also of the body-wall, cause, especially during compression, marked changes in the position of the various organs. It is therefore possible that Schmidt's figure (34, pl. iv, fig. 3) may not represent the natural relations. v. Graff (53) does not mention the position of the ovaries, upon which Schmidt laid stress. For the present, therefore, and until more specimens are available, I place the Plymouth specimen under Schmidt's species, with which in almost all other points it appears to be identical.

Sub-order 2.—TRICLADIDA.*

* This sketch of the two marine Triclads of our shores will at least serve to show how much still remains to be done in the group. The synonymy is very difficult, and requires a thorough revision.

Family PLANARIIDÆ.

Genus 25.—GUNDA, O. Schmidt (1860).

58. GUNDA ULVÆ (Oersted).

? 1768.	HIRUDO LITTORALIS, *Strom* (1).	
? 1776.	PLANARIA LITTORALIS, *Müller* (3).	
1844.	,, ULVÆ, *Oersted* (16).	
1857-8.	PROCERODES ULVÆ, *Stimpson* (32A)	
? 1860.	PLANARIA LITTORALIS, *van Beneden* (33).	
1861.	FOVIA LITTORALIS, *Diesing*, S. B., 'Akad. wiss. Wien,' Bd. xliv.	
1861.	PROCERODES ULVÆ, *Diesing*, loc. cit.	
1865.	PLANARIA ULVÆ, *Johnston* (38).	
1870.	,, ,, *Uljanin* (41).	
1878.	PROCERODES ULVÆ, *Jensen* (49).	
1880.	PLANARIA ULVÆ, *Czerniavsky* (52).	
? 1880.	SYNHAGA AURICULATA, *Czerniavsky* (52).	
1881.	GUNDA ULVÆ, *Lang*, 'Naples Mittheil.,' ii.	
1887.	,, ,, *Iijima*, 'Journ. Coll. Sci. Imp. Univ. Japan,' vol. i, part 4.	
1889.	,, ,, *Wendt*, 'Archiv f. Naturgeschichte,' Bd. i, Heft. 1.	

Length 5—7 mm. *Breadth* ·4—1 mm. *Body* of uniform breadth. Anterior margin truncate, produced at the angles into a pair of distinct forwardly-directed auricles. Behind these a slight "neck" occurs. Posterior end broadly rounded or bifid. *Colour* variable. Young specimens are pale grey. In older examples the pigment is darker, and has a streaky appearance. On the dorsal surface just behind the eyes the pigment is arranged as a median and two lateral bands. The two latter cease behind the "neck." The median one runs forward between the eyes, and then dividing into 3—4 bands, it disappears. The modified strip of integument ('Tast-Organ,' which occurs among all groups of Turbellaria) is present. *Pharynx* inserted at the centre of the body. Two eyes are present anteriorly. They are placed in the white areas bounded by the pigment stripes. The *genital pore* lies slightly behind the commencement of the posterior third. *Testes* occur between the intestinal branches throughout the length of the body. The single pair of *germaria* are placed just behind the eyes and outside the lateral nerves. *Vasa deferentia* run at the sides of the pharynx and open into the peg-like *penis*, which is directed obliquely dorso-ventrally. The *yolk-glands* lie in the septa under the alimentary

canal. The *oviducts unite,* and the common duct thus formed opens
into the neck of the uterus, which is placed behind the genital atrium.
The movements of this animal closely resemble the leech-like progression of fresh-water Planarians.*

HABITAT.—Among roots of Laminaria, Berwick Bay (Johnston, 49) ;
in brackish water on west coast of Scotland (McIntosh, 45).

DISTRIBUTION.—Coast of Denmark, Holland, Belgium, Norway,
Sweden, Black Sea, Baltic.

Genus 26.—FOVIA, Stimpson (32A).

59. FOVIA AFFINIS, Stimpson.   Pl. X, fig. 9.

      1841.   PLANARIA AFFINIS, *Oersted* (16).
      1853.   ,,   HEBES, *Dalyell* (29).
      1857-8. FOVIA AFFINIS, *Stimpson* (32A).
      1865.   PLANARIA AFFINIS, *Johnston* (38).
      1878.   FOVIA AFFINIS, *Jensen* (49).

Length 4—6·5 mm. *Body* linear-oblong, convex above, flat beneath.
The form of the anterior end is described by Johnston and figured
by Dalyell as slightly enlarged and rounded. Oersted's (16) pl. i,
fig. 6, probably represents this species. The explanation of the plate
states it to be *Planaria littoralis,* which, however, is not the case, since
the latter is synonymous with *Planaria ulvæ.* A specimen taken at
Plymouth is figured on Pl. X, fig. 9. The anterior end tapers
slightly, and when viewed "end on" presents two slight lobes, which
are used in a vigorous sensitive way, as in the case of *Convoluta
paradoxa.*

The *colour* varies from greenish-brown to wood-brown. An oval
white spot in the hinder half of the body marks the pharynx. The
two eyes lie each at the inner side of a white area, and from them a
pair of dark parallel streaks of pigment run to the anterior margin.

The *movements* of the animal are very striking. The most usual
method of locomotion is by arching the body and drawing the hinder
end up to the anterior one. These "geometer" or leech-like movements are repeated with great rapidity. This kind of motion is
chiefly effected on moist surfaces. When, however, the water is

* This account is chiefly taken from Iijima, loc. cit.

deeper, the ⟨⟩ gliding ciliary movement is adopted. The hinder part of the ⟨⟩ is kept on the substratum, while the anterior extremity is raised up and constantly extended and retracted, the body as a whole partaking of the steady forward movement.*

HABITAT.—Among algæ, Firth of Forth (Dalyell); Plymouth (F. W. G.).

DISTRIBUTION.—Coasts of Denmark, Norway, and Sweden.

Sub-order 3.– POLYCLADIDA.

A. ACOTYLEA.

Family PLANOCERIDÆ.

* *Mouth and pharynx subcentral. Main-gut rarely extends in front of or behind the pharyngeal sheath. Dorsal tentacles present. Eyes occur (1) on or round the bases of the tentacles; (2) as a double cephalic group; (3) on the body margin. Development usually direct.*

Genus 27.—PLANOCERA, de Blainville,† 1826.

Body broad, leaf-like. Tentacles tapering, contracting into temporary pits. Brain and tentacles lie at the beginning of the second fourth of the body. Marginal eyes absent. Pharynx at rest lies completely folded in its sheath. Two genital apertures some distance from the hinder end.

60. PLANOCERA FOLIUM (Grube).

1840. STYLOCHUS FOLIUM, *Grube* (14).
1844. PLANOCERA FOLIUM, *Oersted* (16).
1856.   ,,   ,,   *Johnston* (38).
1884.   ,,   ,,   *Lang* (54).

* Bergendal ("Studien u. nordischen Turbellarien," 'Ofvers af Kongl. Vetensk Akad. Forhandlingar,' 1890, No. 6) has described a species apparently synonymous with the present one, in which the uterus has a separate external opening. He defines a new genus, *Uteriporus*, containing the single species *U. vulgaris*, Berg. An accident prevented a re-examination of my specimens.

* The definitions of families and genera of Polyclads are taken from Lang (54).

† de Blainville, 'Dictionnaire des Sciences naturelles,' art. "Planaire," t. xli. 1826.

Length 1·4 mm. *Body* extremely contractile, so that a definite shape cannot be stated. The *ground-colour* is pale-yellow ; the main-gut and its branches are brown, ending in marginal black spots. Small white spots (due to the underlying ovaries) are dotted over the dorsal surface. On this surface, nearly one fourth the length of the body from the anterior end, are the cylindrical *tentacles*, which are during life capable of being suddenly extended and as quickly retracted into pits. Round the bases of these tentacles are clusters of eyes. The two *genital apertures* lie behind the mouth, the male pore in front of the female.

HABITAT.—The coralline region, Berwick Bay (Johnston).

DISTRIBUTION.—Palermo (Grube).

Genus 28.—STYLOCHOPLANA, Stimpson (32A).

Planoceridæ with a delicate body expanded anteriorly. Marginal eyes absent. Pharynx only slightly folded at rest. 6—7 pairs of secondary gut-branches. Genital apertures separate or united. Penis unarmed. Penial sheath serves as genital atrium. Vesicula seminalis opens into the vesicula granulorum, and this direct into the ductus ejaculatorius. Bursa copulatrix and accessory vesicle present.

61. STYLOCHOPLANA MACULATA, Quatrefages (18).

? 1836.	PLANARIA SUBAURICULATA, *Johnston* (12).		
1845.	STYLOCHUS MACULATUS, *Quatrefages* (18).		
? 1853.	PLANARIA CORNICULATA, *Dalyell* (29).		
1863.	STYLOCHUS MACULATUS, *Claparède* (36).		
? 1865.	LEPTOPLANA SUBAURICULATA, *Johnston* (38).		
1866.	,,	,,	*Ray Lankester* (39).
1874.	,,	,,	*McIntosh* (45).
1884.	STYLOCHOPLANA MACULATA, *Lang* (51).		

Length 12—16 mm. *Body* flat, elongate, increasing in width from the hinder end forwards, the anterior fourth expanded laterally. The general colour is warm brown, due to the underlying gut-branches. Along the mid-dorsal surface, pale areas indicate the position of the pharynx and the genital pores. A pair of dorsal tentacles are present. *Mouth* mid-ventral, leading into the *pharynx*, the walls of which are plaited. 5—6 eyes are borne by each tentacle, and 7—8 occur round their bases. The *male genital pore* lies at the commencement of the hinder fourth of the length of the body ; the *female pore* a short distance behind this. The penis is pyriform, and receives the contents of a glandular mass distad upon its cord. The *uterus* lies in front and at the sides of the *pharynx*; it opens to the exterior along with the *spermotheca* through the *vagina*.

HABITAT.—Berwick Bay (Johnston); Firth of Forth (Dalyell); Firman Bay, Guernsey (Lankester); St. Andrews (McIntosh); Jersey (Koehler).

DISTRIBUTION.—St. Malo (Quatrefages), St. Vaaste-la-Hogue (Claparède).

Family LEPTOPLANIDÆ.

Mouth and pharynx sub-central. Main-gut usually extends in front of, rarely behind, the pharyngeal sheath. Branches numerous. Male copulatory organ directed posteriorly. Tentacles absent. Eyes— (1) two lateral groups on the areas representing the tentacles of Planoceridæ; (2) double cephalic group; (3) marginal; (4) irregularly disposed over the head. Direct development.

Genus 29.—LEPTOPLANA, Ehrenberg (10).

Leptoplanidæ with elongate, delicate body. Pharyngeal sheath long. Lateral pouches numerous. Pharynx not completely lobed. Genital apertures distinct. Granule-gland and vesicula seminalis separate. Marginal eyes absent. Eyes of group (1) larger than those of (2).

62. LEPTOPLANA TREMELLARIS (O. F. Müller).

1774. FASCIOLA TREMELLARIS, *O. F. Müller* (2).
1776. PLANARIA TREMELLARIS, *O. F. Müller* (4).
1814.   ,,   FLEXILIS, *Dalyell* (6).
1840.   ,,   TREMELLARIS, *W. Thompson* (15).
1844. LEPTOPLANA TREMELLARIS, *Oersted* (16).
1845. PLANARIA FLEXILIS, *Johnston* (20).
1845. POLYCELIS LÆVIGATUS, *Quatrefages* (18).
1849. PLANARIA FLEXILIS, *Thompson* (25).
1853.   ,,   ,,   *Dalyell* (29).
1865. LEPTOPLANA TREMELLARIS, *Johnston* (38).
1866.   ,,   FLEXILIS, *Ray Lankester* (39).
1873.   ,,   ,,   *McIntosh* (43).
1886.   ,,   TREMELLARIS, *Koehler* (55).
1886. POLYCELIS LÆVIGATUS *Koehler* (55).

Length 12—25 mm. *Body* delicate, of variable shape, more or less
elongate, broader in front than behind, the anterior margin semicircular;
young specimens, as Dalyell has remarked, have the outline of a
spherical triangle. The colour, if present, is brown; it is, however,
extremely variable in amount and intensity: it is due partly to
parenchymatous pigment, partly to the gut branches. In the middle
line, not far from the hinder end, are two white areas; the foremost
represents the male copulatory organ, the one behind it the "*shell-
gland.*" Between these and the brain a brown median area, surrounded
by a clear whitish space, represents the main-gut and the uterus outside
it. A **V**-shaped spot leading to the male pore is due to the underlying
vasa deferentia. The ovaries appear (especially on a black ground) as
white dots. From the white ventral surface the plaited pharynx and
genital organs may be seen. Active swimming movements are produced by the expanded edges of the anterior end of the body.
Between the male and female genital apertures is a depression, the
lips of which are strongly muscular, and constitute a "sucker." The
mouth is in front of the centre. It leads into the strongly puckered
pharynx lying in its sheath. From this the main-gut arises, and runs
forwards towards the brain, and backwards to the commencement
of the posterior third, giving off as it does so the numerous lateral
branches, which in turn subdivide and end in fine cæca along the
margin. The *brain* is distinctly bilobed. The lobes are oval, their
long axes parallel to one another and to that of the body. Five
anterior pairs of nerves supply the region in front and at the sides of
the brain, and two lateral ones the rest of the body. The *eyes* vary in
number and arrangement. In front of the brain are usually two
distinct patches of loosely arranged eyes at the bases of the nerves
(corresponding to the eyes at the bases of the tentacles in
Planoceridæ). Opposite the bases of the fifth pair of nerves is a
compact group of larger, chiefly reniform eyes. In some specimens,
however, the tentacular and cephalic groups of each side are continuous
with one another. The genital apertures have already been noticed.
From the numerous scattered *testes*, *vasa efferentia* arise. These
gradually unite to form the pair of *vasa deferentia* which run at the
sides of the pharynx, and before uniting at the base of the penis give
off a posterior branch, which joins the one of the other side behind

the female genital pore. The male copulatory organ consists of a ductus ejaculatorius, and the strongly muscular vesiculæ seminalis and granulorum. The ova scattered throughout the body accumulate after fertilisation in the long uterus, which completely (in adult specimens) surrounds the pharynx and genital-apparatus. The uterus communicates with the exterior by a median duct, which in its lower portion is surrounded by the voluminous "shell-gland."

HABITAT.—Firth of Forth (Dalyell); Cultra, Belfast Bay (W. Thompson); Rothesay (Johnston); Firman Bay, Guernsey (Ray Lankester); St. Andrews (McIntosh); Jersey, Guernsey, Herm (Koehler); Plymouth Sound and neighbourhood, from littoral zone to 20 fms. (W. Garstang, F. W. G.); Hilbre Island, mouth of the Dee, Port Erin, Isle of Man (F. W. G.); Aberystwyth (J. H. Salter).

DISTRIBUTION.—Black Sea, Mediterranean, west coast of France, coast of Holland, Denmark, Baltic, North Sea, Red Sea.

The distinctive peculiarities of this species are the possession of a "sucker" and the simplicity of the female copulatory organ. Thus the "antrum femininum," or cavity into which the female genital pore leads directly, remains simple, while in *Leptoplana alcinoi* and *vitrea* its walls are very muscular, and the organ becomes a bursa copulatrix. External form and colour, as Lang has forcibly stated (54, p. 482), afford no secure basis for the foundation of characters by which the species of Leptoplana may be distinguished.

63. LEPTOPLANA MERTENSII (Claparède).

1861. CENTROSTOMUM MERTENSII, *Claparède* (35).

Length 18 mm. *Body* oval, white, or yellowish. Two groups of eyes on the dorsal surface.

As no details as to the structure of the genital organs are given, it is impossible to satisfactorily assign a position to this species.

HABITAT.—On Laminaria, Lamlash Bay, Arran (Claparède).

APPENDIX TO LEPTOPLANIDÆ.

64. PLANARIA ATOMATA. O. F. Müller.

1777. PLANARIA ATOMATA, *O. F. Müller* (4).
1823. „   „   *Fleming* (8).
1839. „   „   *Forbes and Goodsir* (13).

1844. LEPTOPLANA ATOMATA, *Oersted* (16).

? 1845.   „   DRŒBACHENSIS, *Oersted* (21).

? 1853. PLANARIA MACULATA, *Dalyell* (29).

1865. LEPTOPLANA ATOMATA, *Johnston* (38).

1874.   „   „   *McIntosh* (45).

Length 10—12 mm. *Body* oval, slightly wider in front than behind, rounded at both extremities. *Colour* variable, the ground-tint white or brown, spotted with reddish-brown, white beneath. *Ova* are seen as white dots over the upper surface. Two "tentacular" and two cephalic groups of *eyes* are present. The only known fact about the genital organs is that the penis has a bulbous base, and a transparent terminal duct which contains a hard stylet.

HABITAT.—Coast of Scotland (Fleming); Orkneys and Shetlands (Forbes and Goodsir); Firth of Forth (Dalyell); St. Andrews (McIntosh).

DISTRIBUTION.—Naples (Delle Chiaje), coast of Holland, Germany, Denmark, Baltic (Dröbaek).

Planaria atomata has never been described in a sufficiently diagnostic way to render possible the identification of specimens with it. Consequently the above synonymy is very probably incorrect, but it is in no one's power to tell what the authors quoted did mean by their *Planaria atomata*. Thus Forbes and Goodsir, Fleming, Johnston, and McIntosh merely give the name and the record. Even those (as Müller and Oersted) who vouchsafe anatomical facts state the size, the form of the body, the position and arrangement of the eyes, and the form of the penis, and these do not by themselves, signalise a species of Leptoplana. Evidently a fresh and full description of a form is needed, which, if it differs from other existing species, may be called *atomata*, although its unity with the species of that name can only extend to the points mentioned. Comparisons with the new fully described *atomata* would henceforth be possible. Such a full account of a species agreeing in the form of the body, the position of the eyes, and the composition of the penis is to be found in Jensen's description of *Leptoplana Drœbachensis*.

The small differences that justified Oersted in separating these two species were the following:

Oersted (21)	*Drøbachensis*, Oe.	*Leptoplana atomata*, O. F. Muller.
	Length 4 lines.	Length 3—4 lines.
	Body "antice obtuso, dein sensim angustiore."	Body "subovali, postice angustiore."
	Eyes arranged in an anterior linear clump, and a posterior triangular one of 7.	Eyes arranged in four clumps. Those of the posterior ones are the larger.

With regard to the arrangement of the eyes, Jensen's specimens of *L. Drøbachensis* differ from Oersted's just as much or as little as does *L. atomata*. Oersted does not mention a hard penis in *Drøbachensis*, although he describes it in *L. atomata*. Taking these facts into account, it may, perhaps, be said that Oersted's and Jensen's specimens have as much right to be classed in the same species as has Oersted's *L. atomata*.

For the future recognition of *L. Drøbachenses* I quote Jensen's diagnosis.

Length 10 mm. *Breadth* 4—5 mm. *Body* slightly narrowed posteriorly, rounded at both ends. Dorsal surface red with scattered darker spots, and with a longitudinal area, down the centre of which runs a broken white line. Ventral surface white. Four paired groups of *eyes*. Anterior groups placed longitudinally, hinder group directed outwards and backwards, composed of larger eye-specks. *Mouth* subcentral. *Penis* styliform, hardened at its apex or for its whole length. *Vagina* connected with the *spermotheca* by a long duct provided with a moniliform series of dilatations (Jensen [49], pl. vii, figs. 10—14).

B. COTYLEA.

Family EURYLEPTIDÆ.

Cotylea usually provided with marginal tentacles. Brain anterior, behind the tentacles. Mouth just behind, rarely in front of the brain. Pharynx directed forwards, cylindrical. Main-gut behind the pharyngeal sheath. Male copulatory organ simple, directed forwards, placed just behind or beneath the sheath of the pharynx. A hard stylet in the penis. Female genital pore between the penis and the sucker. Eyes (1) in or round the tentacles; (2) double cephalic group, sometimes greatly elongated.

Q

Genus 30.—Prosthecer.eus, Schmarda.*

Body smooth, delicate. Pharynx bell-shaped. Main-gut extending to posterior extremity. Body around the pharynx and main-gut frequently thickened. Uterine glands corresponding to the number of the secondary gut-branches. Tentacles well developed, pointed, moveable. Two small cephalic groups of eyes. Brightly coloured forms.

65. Prosthecer.eus vittatus (Montagu).

1815.	Planaria vittata,	*Montagu* (7).	
1823.	,,	,,	*Fleming* (8).
1840.	,,	,,	*Thompson* (15).
1845.	,,	,,	*Johnston* (20).
1846.	,,	,,	*Thompson* (22).
1857.	,,	,,	*Harvey* (31).
1865.	Eurylepta vittata,	*Johnston* (38).	
1884.	Prosthecer.eus vittatus,	*Lang* (54).	
1886.	,,	,,	*Koehler* (55).

Length 3·7—5 cm. *Body* elliptical, tapering towards both extremities. The *tentacles* are lamellar, broad at their bases, which enclose between them the extreme anterior tip of the body in such a way as to separate it off slightly from the margin. The general *colour* is yellow, the margins white. The median ridge is distinguished by a black line. Right and left of this a number of black lines (increasing in number and distinctness with the age of the individual) run from the brain towards the hinder end. Those near the median plane are almost straight; the peripheral run parallel with the body margin. The *mouth* lies behind the brain. The *main-gut* is long, and gives off large numbers of secondary gut branches, which anastomose freely. *Eyes* occur over the brain as a pair of small, clearly-defined cephalic groups on the anterior margin and in the tentacles. The *sucker* is subcentral. Halfway between it and the mouth is the *male aperture*, and behind this the *female genital pore*. The *vesicula seminalis* and *granule-gland* open independently into the *ductus ejaculatorius*.

Habitat—Estuary of Kingsbridge, S. Devon (Montagu); coast of Scotland (Fleming); Strangford Lough (W. Thompson); between tide-marks at Roundstone, Connemara (W. Thompson); British coast

* Schmarda, L. K., " Neue wirbellose Thiere, beobachtet und gesammelt auf einer Reise um die Erde, 1853—1858,' Bd., i, 1859.

(Harvey) ; Falmouth (J. Cranch, vide Johnston, 38) ; Jersey, Guernsey, Herm (Koehler) ; two specimens off Stoke Point, near Plymouth, 15 fms., on *Diazona* (J. T. Cunningham, MSS.) ; Plymouth Sound (W. Garstang).

DISTRIBUTION.—Mediterranean, west coast of France, Scandinavia, Denmark.

66. PROSTHECERÆUS ARGUS (Quatrefages).

            PROCEROS ARGUS,

            PROSTHECERÆUS   cit.

            EURYLEPTA ARGUS, *Keferstein* (40).

            PROSTHECERÆUS ARGUS, *Lang* (54).

      1886.   PROCEROS ARGUS, *Koehler* (55).

Length 6—10 mm. *Body* oval, bearing two short marginal tentacles separated by the anterior extremity. *Dorsal surface* somewhat convex, orange with white spots. The *eyes* are numerous : on each side of the middle line extending behind the brain, and continued forwards to the ventral faces of the tentacles. Thus the marginal, tentacular, and cephalic groups are continuous.

HABITAT.—Between tide-marks at Grand Havre, Guernsey (Koehler).

DISTRIBUTION.—St. Malo (Quatrefages, Keferstein).

Genus 31.—CYCLOPORUS, Lang (1884).

Dorsal surface papillose. Pharynx short. Cells of main-gut almost ... About ... pairs of secondary branches. Uterine glands ... to the ... of the latter. The peripheral gut-branches open to the exterior through epithelial pores. Male pore close behind the mouth. Copulatory organ beneath and behind the pharyngeal sheath. Cephalic group of eyes not sharply defined. Tentacles small, sometimes rudimentary.

67. CYCLOPORUS PAPILLOSUS, Lang (54). Pl. X, fig. 2.

      1889.   PROCEROS TUBERCULATUS, *Schmidtlein*, 'Mittheil. Zool. Stat. Stat. Neapel,' Bd. ii.

      1884   ,,   ,,   *Lang.* ibid., Bd. iii.

Length 10—14 mm. *Body* elliptical with blunt extremities. In front the antero-lateral margins are produced to a variable extent as a pair of small, pointed tentacles. The *dorsal surface* is typically covered with small coloured papillæ—absent, however, in the variety *lævigatus*. Excepting the margins, the body is opaque. The *ground-colour* is yellowish-white. The main-gut and its six pairs of branches are brown, red, yellow, &c. In adult specimens they are largely concealed by the genital organs, but reappear on the margin, where their terminations are usually brightly coloured. The colour of the dorsal tubercles is variable and due to pigment in the epidermis. When the tubercles are absent their position is indicated by pigment-spots. Thus the colour is due partly to the contents of the gut, to pigment, and to the genital organs. Combinations of these three sources of colour account for the diversity between individuals of the same and of different ages, and appear to be correlated with the substratum (generally species of *Leptoclinum* and other Ascidians). Three to four black spots are present in specimens of the *lævigatus* variety, round the first pair of secondary gut-branches. (For good descriptions of the appearance of this animal at different stages of growth see Lang, pp. 54, 568—571).

The mouth, just behind the brain, leads into a conical *pharynx*, the apex of which is directed backwards and is continued into the long *main-gut*. From this six lateral pairs of branches arise at right angles, which, after branching and anastomosing freely, end in terminal vesicles opening to the exterior through temporary *epidermal pores* at the moment of the expulsion of fæcal matter. The elongate cephalic group of *eyes* borders the white area produced by the pharynx, and extends forward beyond the brain. There is also a distinct group at the base and on the ventral surface of the tentacles. The *male genital pore* lies close behind the mouth, the *female aperture* halfway between the anterior end and the subcentral sucker. The *vesicula seminalis* is very large. The *uterus* is a large lobed sac surrounding the main gut; the *uterine glands* are numerous (10–11 on each side). Surrounding the female pore is the very large radiate "*shell-gland.*"

HABITAT.—On compound Ascidians and the sponge *Hymeniacidon sanguinea*, 5—15 fms., Plymouth (W. Garstang, F. W. G.); Port Erin, Isle of Man, 18 fms. (H. C. Chadwick). Var. *lævigatus* between tide-marks, Port Erin, and neighbourhood (W. J. Beaumont, F. W. G.).

DISTRIBUTION.—Naples (Lang).

This variable species may be easily mistaken for *Stylostomum variabile*. It is, however, recognisable by the presence of a continuous median gut-branch over the pharynx, whereas in *Stylostomum* the pharyngeal region appears as an uninterrupted white area, bordered laterally by the gut-diverticula.

Genus 32.—EURYLEPTA, Ehrenberg, 1831 (10).

Body smooth. Tentacles long, tapering. Usually five pairs of secondary gut-diverticula. The intestine is brightly coloured. Male genital pore beneath the hinder end of the pharyngeal sheath. One pair of uterine glands is present. Cephalic group of eyes extending posteriorly far beyond the brain.

68. EURYLEPTA CORNUTA (O. F. Müller).

1776.	PLANARIA CORNUTA, *O. F. Müller* (3).		
1831.	EURYLEPTA CORNUTA, *Ehrenberg* (10).		
1832.	PLANARIA CORNUTA, *Johnston* (11).		
1845.	,,	,,	*Thompson* (19).
1845.	,,	,,	*Johnston* (20).
1853.	,,	,,	*Dalyell* (29).
1865.	EURYLEPTA CORNUTA, *Johnston* (38).		
1865.	,,	DALYELLI, *Johnston* (38).	
1866.	,,	CORNUTA, *Ray Lankester* (39).	

Length 1·5—3·75 cm. *Breadth* about half the length. *Body* elliptical during motion, almost circular at rest, broadly rounded behind. In front are two elongate tentacles. The somewhat convex dorsal surface is, with the exception of the margins, opaque, of a bright orange-red colour dotted with white, due to parenchymatous pigment, and to a greater extent to the contents of the alimentary canal. In front an elongated, oval, raised, white ridge represents the underlying pharynx. The ventral surface is of a pale reddish colour, and upon it the gut, male and female apertures, and sucker are visible. The *mouth* is one-third of the distance from the anterior end to the sucker, *i.e.*, close behind the brain. The *pharynx* is well developed, cylindrical, extending almost as far back as the centre, in front of which it opens into the extensive *main-gut*. From this a median and 5—6 lateral pairs of branches arise, which after branching slightly end in marginal forked cæca. The minutely moniliform appearance of these is due to the presence of sphincter-muscles at the points of constriction. Each tentacle receives a branch of the intestine. Extending from the brain towards the hinder margin of the pharynx are two groups of *eyes*, which stand out very clearly against the white underlying pharyngeal region. Posteriorly they are divergent, and consist of small, loosely aggregated eye-specks, which become larger and crowded in front, the two groups converging towards the brain. Eyes are also present in the tentacles and around their bases. The *sucker* is well developed, and serves to attach the animal very firmly to the substratum. The *male genital aperture* lies under the hinder end of the pharynx. The *vasa efferentia* unite in a large expanded duct on each side, from which the two *vasa deferentia* arise. The *ductus ejaculatorius* receives the contents of the *granule-gland* and the large vesicula seminalis. Just in front of the sucker lies the *female pore*, surrounded by the extensive radiating *shell-gland*. The *uterus* lies at the sides of the main-gut. A single pair of *uterine glands* are present.

HABITAT.—" On the coast of Berwickshire, in deep water on corallines and shells " (Johnston, 11); on Laminaria, 6—10 fms., Belfast Bay (W. Thompson); Firth of Forth (Dalyell); Firman Bay, Guernsey (Ray Lankester); Bordeaux (Koehler); Plymouth, in 2—6 fms., and between tide-marks (W. Garstang F. W. G.).

DISTRIBUTION.—Naples (var. *Melobesiarum*, Lang), St. Malo (Keferstein), Dröback (Müller).

Genus 33.—OLIGOCLADUS, Lang (1884).

Body smooth. Tentacles long, capable of movement. Mouth in front of the brain. Pharyngeal sheath produced posteriorly into a closed diverticulum, which extends beyond the sucker. Pharynx cylindrical. Three to four pairs of secondary gut-branches. The main-gut apparently opens to the exterior at its hinder end. Male and female genital apertures lie under the pharyngeal sheath. Four pairs of uterine glands are present. The double cephalic eye-group sharply defined, not elongated behind.

69. OLIGOCLADUS SANGUINOLENTUS (Quatrefages). Pl. X, fig. 3.

1845. PROCEROS SANGUINOLENTUS, *Quatrefages* (18).
? 1864.   „   „   *Grube**
  * 'Die Insel Lussin u ihre Meeresfauna,' 1864.
1884. OLIGOCLADUS SANGUINOLENTUS, *Lang* (54).
1886.   „   „   *Koehler* (55).

Length 8—11 mm. *Breadth* 3—4 mm. *Body* delicate, fairly transparent. *Form* elongate, parallel-sided, broadly rounded behind, produced in front into a pair of long, pointed, contractile tentacles, between which the extreme anterior end projects slightly. *Colour* white, especially marked along the margins. The mid-dorsal line is brownish or carmine, owing to the underlying *main-gut* and its median branch. The latter exhibits two conspicuous swellings, one at the point of origin, the other behind the brain. Three to four lateral diverticula arise on each side of the main-gut, and are of a brilliant carmine colour at first, becoming much less conspicuous towards the periphery. The pharynx and genital organs appear as white patches round the main-gut. The *mouth* is placed in front of the brain. The gut-branches do not anastomose. *Eyes* are present at the bases of the tentacles, and two sharply defined cephalic groups converge at the anterior end of the brain. The position of the genital apertures has already been mentioned. The male pore lies in front of the female.

HABITAT.—Between tide-marks, Grève d'Azette, Jersey (Kohler); Plymouth Sound, 5—20 fms. (F. W. G.); Port Erin, Isle of Man, 12—15 fms. (W. J. Beaumont and F. W. G.).

DISTRIBUTION.—Island of Lussin, Adriatic (Grube), Naples (Lang).

After much consideration I have referred several specimens dredged at Plymouth and elsewhere to this species. The distinguishing points are the position of the mouth in front of the brain; the male genital aperture underneath the anterior end of the pharyngeal sheath; and the short, sharply defined group of eyes over the brain.

70. OLIGOCLADUS AURITUS (Claparède).

1861.   EURYLEPTA AURITA, *Claparède* (35).
1884.   OLIGOCLADUS AURITUS, *Lang* (54).

Length 18·5 mm. *Body* oval, transparent, white, the intestine bright reddish brown. *Mouth* in front of the brain. *Pharynx* cylindrical. *Main-gut* gives rise to three pairs of secondary branches, which do not anastomose. *Eyes* are present in and round the bases of the tentacles, but, according to Claparède, are absent over the brain. The *male genital pore* occurs just behind the mouth; the *female aperture* is described by Claparède as almost central. Lang suggests that this author probably mistook the sucker for the pore. The *vasa deferentia* are scarcely so swollen as in *O. sanguinolentus*, and the *vesicula seminalis* rather larger than in the latter. The *granule gland* and *copulatory organ* agree exactly in both species. Claparède figures large rounded bodies which may possibly prove to be the accessory uterine glands (Lang).

HABITAT.—On Laminaria, Lamlash Bay, Arran (Claparède).

A more exact description of this species is necessary before the specific identity or difference of *Oligocladus sanguinolentus* and *auritus* can be regarded as proved. It appears fairly clear that they both possess the same generic characters—the subterminal mouth, position of the genital pores, and multiple uterine glands. No satisfactory points of difference can at present be determined. On the contrary, it is noticeable that it is just those organs which Claparède describes accurately—the mouth, pharynx, and intestine, and the male copulatory organ—which agree exactly with the corresponding structures in *O. sanguinolentus*.

Genus 34.—STYLOSTOMUM, Lang (1884).

Body smooth. Tentacles rudimentary. Oral and genital apertures open on a common depression immediately behind the brain. Maingut with 5—6 pairs of secondary non-anastomosing branches. The median anterior branch is absent over the pharyngeal region. Male copulatory organ lies under the anterior part, the female organ under and behind the hinder part of the pharyngeal sheath. Two uterine glands. Cephalic eyes few in number.

71. STYLOSTOMUM VARIABILE, Lang. Pl. X, fig. 1.

? 1853. PLANARIA ELLIPSIS, *Dalyell* (29).
? 1865. LEPTOPLANA ELLIPSIS, *Johnston* (38).
? 1875. „ „ *McIntosh* (45).
1884. STYLOSTOMUM VARIABILE, *Lang* (54).
1884. STYLOSTOMUM? ELLIPSIS, *Lang* (54).

Length 5—9 mm. *Body* elliptical, broadly rounded behind, tapering slightly in front. The extreme anterior margin truncate. *Tentacles* more conspicuous in adults than in young specimens, where they form mere blunt marginal projections. Immature specimens derive their coloration from the white or yellowish-white ground-tint and from the branches of the intestine, which, owing to the transparency of the body, are clearly visible. The colour of the gut-branches is scarcely the same in any two specimens, and may be red, orange, brown, black, &c. In mature examples the genital organs conceal the greater part of the alimentary canal. The *mouth* lies immediately behind the brain. It leads into a cylindrical *pharynx*, which, lying in its sheath, appears from the dorsal surface as a white oval area. Bounding the sides of this are the first pair of gut-branches, a median branch being absent. In front of the pharynx these two branches unite and from this point a very short median branch runs to the anterior end. *Eyes* are present below and above the tentacle bases, and also as two divergent series over and slightly beyond the brain. Very characteristic are two pairs of eyes close to the hinder margin of the brain, and a pair on its outer and anterior angles. The relation of these eyes to those of the larva may be gathered from Pl. X, fig. 1, which represents a young specimen of the present species. The *male genital pore* is combined with the mouth behind the brain; the *female pore* lies in front of the centre, the sucker just behind it. *Granule-gland* and *vesicula seminalis* open into the penis. The vesicula receives the separate *vasa deferentia*. The *uterus* encloses the main-gut. A very extensive *shell-gland* surrounds the female genital pore.

HABITAT.—Firth of Forth (Dalyell, 29); not uncommon between tide-marks (McIntosh 45); Falmouth, at low water (W. Garstang); Plymouth, in 4½ fms., along with young specimens; Port Erin, Isle of Man, in 12 fms. (F. W. G.).

DISTRIBUTION.—Naples (Lang).

This species, closely similar to young smooth specimens of *Cycloporus papillosus*, may be distinguished by the absence of a median gut-branch over the white pharyngeal region, by the presence of only 5—6 pairs of secondary branches (Cycloporus possesses 8—9) to the intestine, and by their non-anastomosing character.

I have included *Planaria ellipsis* of Dalyell and others under this species, since his figures agree exactly in the points just mentioned.

III. SUMMARY.

1. British marine Turbellaria, as at present known, include about fifty-seven species of Rhabdocoelida, twelve of Polycladida, and two

Tricladida, making a total of seventy-one species.   The numbers represent the examination of a limited extent of our coast (Millport, St. Andrews, Skye, the Isle of Man, Plymouth, and Channel Islands) during about three months of the year (July to September).

2. The following twenty-eight species are added to the British fauna in the present paper.

POLYCLADIDA :

Cycloporus papillosus, Lang.

ACŒLA :

Proporus venenosus (O. Sch.).
Monoporus rubropunctatus (O. Sch.).
Aphanostoma elegans, Jensen.

RHABDOCŒLA :

Promesostoma ovoideum (O. Sch.).
   ,,   *solea* (O. Sch.).
   ,,   *agile* (Lev.).
   ,,   *lenticulatum* (O. Sch.).
Byrsophlebs Graffi, Jensen.
Mesostoma neapolitanum, v. Graff.
Hyporhynchus armatus (Jensen).
?   ,,   *penicillatus* (O. Sch.).
Provortex rubrobacillus, n. sp.

ALLŒOCŒLA :

Plagiostoma dioicum, Metschnff.
   ,,   *sulphureum*, v. Graff.
   ,,   *Girardi* (O. Sch.).
   ,,   *pseudomaculatum*, n. sp.
   ,,   *sagitta* (Vlj.).
   ,,   *elongatum*, n. sp.
   ,,   *caudatum*, Lev.
   ,,   (?) *siphonophorum* (O. Sch.).
Vorticeros luteum, v. Graff.
Cylindrostoma inerme, Halley.
   ,,   *elongatum*, Lev.
Monoophorum striatum (v. Graff).

Monotus albus, Lev.

Automolus horridus, n. sp.

    ,,   *? ophiocephalus* (O. Sch.).

3. The relations of the Turbellarian fauna of our coasts with that of neighbouring seas cannot be determined with certainty until more extended observations are recorded than we possess at present. Mediterranean and Adriatic forms occur on our south-western stations (Plymouth, &c.). Thus seven Polyclads and sixteen Rhabdocœles (33 per cent. of our fauna) are common to Naples, Trieste, and Plymouth. A large proportion (about 70 per cent.) of Scandinavian forms occur on our coast.

IV. Appendix.

Synopsis of the Families, Sub-families, Genera, and Species of British Marine Turbellaria.

I. RHABDOCŒLIDA.

Section A.—Acœla.

1. With a single genital pore · · · Family **Proporidæ**.

   a. Without spermotheca - · · · Genus Proporus.

      Species :—P. venenosus (elongate, yellow, two eyes present, provided each with a large lens).

   b. With spermotheca · · · Genus Monoporus.

      Species :—M. rubropunctatus (eyes without lenses, composed of red pigment-masses placed in the epidermis).

2. With two genital pores, the female pore in front of the male

                              Family **Aphanostomidæ**.

   a. Spermotheca with soft, non-chitinous mouth-piece

                              Genus Aphanostoma.

      Species :—A. diversicolor (central part of the anterior end violet, extremities yellow). *A. elegans* (centre of the body with a lobate green spot).

   b. Spermotheca with chitinous mouth-piece   Genus Convoluta.

      Species :—C. saliens (with alternate longitudinal rows of cilia and rhabdites). *C. paradoxa* (with two eyes and "yellow cells" in the parenchyma). *C. flavibacillum* (dorsal surface very convex, no "yellow cells," sides of the body only slightly flexed ventrally).

Section B.—RHABDOCŒLA.

3. With sexual and asexual reproduction. Female accessory organs absent - - - - - - - Family **Microstomidæ**.

    a. Sexes separate. Head with a pair of lateral grooves. A pre-œsophageal cæcum present - - Genus MICROSTOMA.

        Species:—*M. grœnlandicum* (eyes absent, a red spot usually present anteriorly).

    b. Hermaphrodite. Proboscis present. A posterior and some-times lateral bundles of setæ only) - Genus ALAURINA.

        Species:—*A. Claparedii* (proboscis with numerous papillæ, and a pair of ciliary tufts at its base; posterior bundle of setæ only).

4. With one or two genital apertures. Male accessory organs present. Pharynx usually mid-ventral, rosulate (i.e. rosette-like)

                                     Family **Mesostomidæ**.

  I. A single genital aperture, two germaria, and two vitellaria. Accessory reproductive organs absent

                             Sub-family **Promesostominæ**.

    a. Characters of sub-family - - Genus PROMESOSTOMA.

        Species:—*P. marmoratum* (copulatory organ coiled, crosier-like). *P. ovoideum* (penis pyriform, pigment-cup of eye simple). *P. solea* (pigment-cup of eye with hook-like process over outer surface of lens). *P. lenticulatum* (copu-latory organ provided distally with radial triangular ridges). *P. agile* (copulatory organ curved, simple).

  II. Two genital apertures. The male pore in front of the female. Germarium single - - - Sub-family **Byrsophlebinæ**.

    a. Characters of sub-family - - Genus BYRSOPHLEBS.

        Species:—*B. Graffi* (vitellaria unbranched; copulatory organ widely funnel-shaped, the terminal margin with a triangular projection). *B. intermedia* (vitellaria branched, copulatory organ elongate, the terminal margin entire, and with a curved chitinous spur).

III. A common genital aperture. Germ-yolk-glands present. Testes
rounded. Spermotheca with chitinous appendages

Sub-family **Proxenetinæ.**

a. Characters of sub-family   -   -   - Genus PROXENETES.

Species :—P. flabellifer (copulatory organ retort-shaped, complex; duct of the spermotheca with chitinous teeth). *P.
cochlear* (copulatory organ composed of three spoon-shaped
pieces).

IV. A common genital pore. One germarium. Female accessory
organs present. Testes elongate - Sub-family **Eumesostominæ.**

a. Copulatory organ traversed throughout its length by the
ducts of male secretions   -   -   - Genus MESOSTOMA.

Species.—M. neapolitanum (copulatory organ funnel-shaped,
the margin with a spur ; atrium very large).

5. The anterior extremity converted into a tactile proboscis, provided with a (usually complex) musculature - Family **Proboscidæ.**

I. Proboscis simple, with a sheath or muscle-cone. Short muscle-
bundles serve as retractors Sub-family **Pseudorhynchinæ.**

a. Characters of sub-family   -   - Genus PSEUDORHYNCHUS.

Species :— P. bifidus (hinder extremity bifid ; copulatory
organ conical, with a screw-like ridge on its outer surface).

II. Proboscis provided with a sheath opening in front, a muscle-
cone, and four long retractors.   . Sub-family **Acrorhynchinæ.**

a. Distinct seminal and granule vesicles, enclosed, however,
in a common muscular sheath   - Genus ACRORHYNCHUS.

Species :—A. caledonicus (copulatory organ composed of small
chitinous spines).

b. Duct of granule-vesicle with special chitinous investment

Genus MACRORHYNCHUS.

Species :—M. Naegelii (copulatory organ tubular, with a curved
spur longer than the tube). *M. croceus* (chitinous tube long,
continued directly into the " spur "). *M. helgolandicus*
(a chitinous investment enveloping granule-vesicle and vas
deferens ; a poison-dart present).

c. Two genital pores, the female in front of the male pore.
Granule-vesicle with chitinous investment Genus GYRATOR

Species:—*G. hermaphroditus* (colourless ; copulatory organ with a straight poison-dart).

III. Proboscis small, behind the anterior end, its sheath opening on the ventral surface. Granule- and seminal vesicles not separate. Their contents, however, issue by distinct ducts

Sub-family **Hyporhynchinæ.**

     a. Characters of the sub-family -   - Genus HYPORHYNCHUS

         Species:—*H. armatus* (copulatory organ composed of two fused chitinous, spiral tubes ; pigment not reticular ; six papillæ round the mouth. *H. penicillatus* (copulatory organ composed of two spoon-shaped pieces).

6. A single genital aperture. Pharynx large, dolioform. A uterus and paired testes present -   -   -   - Family **Vorticidæ.**

     I. Germaria small, body-cavity capacious, free-living

Sub-family **Euvorticinæ.**

         a. Two germaria and two unbranched vitellaria

Genus PROVORTEX.

         Species:—*P. balticus* (copulatory organ with a spirally-curved spur on the margin). *P. affinis* (copulatory organ slightly bent distally). *P. rubrobacillus* (with red rods in the gut-cells ; copulatory organ with a finely pointed straight spur to the margin).

Section C.—ALLŒOCŒLA.

7. An otolith absent -   -   -   - Family **Plagiostomidæ.**

     I. Genital aperture single, ventral, posterior. Mouth anterior. Germaria present. -   -   - Sub-family **Plagiostominæ.**

         a. Without tentacles -   -   -   - Genus PLAGIOSTOMA.

         Species:—A. With four distinct eyes : *Pl. sagitta.*

            B. With two eyes.

            AA. Mouth terminal or subterminal : *Pl. dioicum* (1—5 mm. long, yellow, pharynx in front of brain). *Pl. elongatum* (white, pharynx large, when retracted it lies behind the brain). *Pl. ochroleucum* (5·5 m.m., pharynx subterminal).

            BB. Pharynx and mouth behind brain.

  α. Epidermis without pigment : *Pl. Girardi* (colourless, 2 reniform eyes). *Pl. siphonophorum* (with median
longitudinal band of black reticular pigment). *Pl. pseudomaculatum* (violet pigment between the eyes, without
distinct lateral grooves). *Pl. vittatum* (pigment variable,
usually in the form of three transverse bands).

  β. Epidermis pigmented : *Pl. sulphureum* (epidermis
with yellow rods). *Pl. Koreni* (transverse band of
parenchymatous pigment). *Pl. caudatum* (rhabdites
few, head marked off by lateral grooves).

 b. With two tentacles at the anterior end - Genus VONTICEROS.
  Species :—V. auriculatum (violet reticular pigment over the
greater part of dorsal surface). *V. luteum* (pigment uniformly yellow).

II. A single posteriorly-placed genital aperture. Two germaria
and two distinct vitellaria. Pharynx directed backwards.

         Sub-family **Allostominæ.**

 a. Without a circular ciliated groove on the head

         Genus ENTEROSTOMA.

  Species :—E. austriacum (four eyes ; pigment in rounded
yellow masses). *E. fingalianum* (pigment absent, colour
due to food). *E. cæcum* (eyes absent).

 b. With a circular ciliated groove at the level of the brain

         Genus ALLOSTOMA.

  Species :—A. pallidum (adults 2—3 mm. long ; granular
mucus-rods [pseudo-rhabdites] in the epidermis).

III. Circular ciliated groove on the head. Oral and genital apertures combined. A pair of germ-yolk-glands present.

         Sub-family **Cylindrostominæ.**

 a. With characters of sub-family - Genus CYLINDROSTOMA.
  Species :— Cyl. quadrioculatum (pharynx directed forwards ;
body colourless). *Cyl. inerme* (epidermis yellow, containing
rhabdites but no " calcareous bodies ;" spermotheca absent).
Cyl. elongatum (pharynx directed backwards ; four eyes
present).

 b. Pharynx directed backwards, the penis forwards. Spermotheca opens into genital atrium. - Genus MONOOPHORUM.

Species :— M. striatum (pigment carmine, reticular ; muscles grouped in longitudinal bundles).

8. With two genital apertures, two germaria, and two vitellaria. An otolith present　-　　-　　-　　- Family **Monotidæ.**

    a. Female genital pore in front of the male　- Genus MONOTUS.

    Species :— M. lineatus (an eye present in front of the otolith ; copulatory organ a soft papilla). *M. fuscus* (copulatory organ a chitinous tube). *M. albus* (penis, a boat-shaped, chitinous copulatory organ with a pair of lateral teeth near the "bows").

    b. Female genital pore behind the male aperture

                       Genus AUTOMOLOS.

    Species :— A. ophiocephalus (an eye usually present in front of the otolith ; head expanded). *A. unipunctatus* (an eye absent ; penis a spinous tube). *A. horridus* (head slightly marked off from the body ; rhabdites giving a spinous appearance to the animal ; penis soft, muscular).

II. TRICLADIDA.

A single (rarely double) genital aperture behind the mouth. Pharynx central or post-central　-　　-　　-　　-　　- Family **Planariidæ.**

    a. Penis directed dorso-ventrally. Uterus opens into genital atrium. Head truncate. Eyes wide apart　Genus GUNDA.

    Species :— G. ulvæ (the angles of the anterior margin produced into "lappets").

    b. Head produced in the centre. Eyes approximated

                       Genus FOVIA.

    Species :— F. affinis (pharynx behind the middle).

III. POLYCLADIDA.

Section A.—ACOTYLEA.

1. Dorsal contractile tentacles present. Tentacular, cephalic, and marginal groups of eyes　-　　- Family **Planoceridæ.**

    a. Body leaf-like. Marginal eyes absent　- Genus PLANOCERA.

    Species :— Pl. folium (tentacles placed behind commencement of second quarter of length ; body yellow-brown).

b. Body distinctly enlarged in front.   Tentacles placed well
apart at end of first fifth of the body

Genus STYLOCHOPLANA.

Species:—St. maculata (two genital apertures).

2. Mouth subcentral.   Main-gut extending in front of (rarely
behind) the pharyngeal region.   Tentacles absent

Family **Leptoplanidæ**.

a. Body slightly enlarged in front.   Marginal eyes absent

Genus LEPTOPLANA.

Species:—L. tremellaris (a "sucker" present between the male
and female genital pores).   *L. Mertensii* and *L. abroata*
(doubtful species, see pp. 239–40).

Section B.—COTYLEA.

3. Anteriorly-placed marginal tentacles usually present.   Mouth
just behind (rarely in front of) the brain.   Penis with a hard stylet

Family **Euryleptidæ**.

a. Brightly coloured.   Body thickened over the main-gut.   Tentacles large, moveable.   Two small cephalic eye-groups.

Genus PROSTHECERÆUS.

Species:—Pr. vittatus (body yellow, with longitudinal thin
black lines).   *Pr. argus* (dorsal surface orange-coloured ;
length up to 10 mm.).

b. Dorsal surface usually papillose.   Tentacles small.   Peripheral
gut-pores to the exterior ·   -   · Genus CYCLOPORUS.

Species:—C. papillosus (a median gut-branch over pharyngeal
region).

c. Intestine brightly coloured.   Tentacles long.   Mouth behind
brain.   Cephalic eye-groups extending far behind brain.

Genus EURYLEPTA.

Species:— E. cornuta (eyes not quite extending back to hinder
edge of white pharyngeal region ; body bright red).

d. Tentacles long.   Mouth in front of brain.   Pharyngeal sheath
with a posterior cæcum   -   · Genus OLIGOCLADUS.

Species:—O. sanguinolentus (body whitish, intestine usually
carmine ; a short cephalic eye-group).   *O. auritus* (eyes
absent over the brain [Claparède]).

R

e. Tentacles rudimentary.   Oral and male genital apertures
united behind brain.   A median gut-branch absent over
the pharyngeal region   -   -   Genus STYLOSTOMUM.
Species:—S. variabile (intestinal branches brightly coloured).

V. LITERATURE REFERRED TO, ARRANGED ACCORDING TO THE DATES OF PUBLICATION.

(1) 1768. STRÖM, H.—"Beskrivelse over Norske Insecter; Andet Stykke," 'Det kongelige Norske Videnskabers Selskabs Skrifter,' Deel iv.

(2) 1773. MÜLLER, O. F.—'Vermium terrestrium et fluviatilium, seu animalium infusoriorum, helminthicorum, et testaceorum non marinorum, succincta historia,' vol. i.

(3) 1776. MÜLLER, O. F.—'Zoologiæ Danicæ prodromus, seu Animalium Daniæ et Norvegiæ indigenarum characteres, &c.,' Havniæ, 8vo.

(4) 1777—1806. MÜLLER, O. F.—'Zoologica Danica,' folio.

(5) 1780. FABRICIUS, O.—'Fauna grœnlandica,' 8vo., pp. 326-7.

(6) 1814. DALYELL, J. G.—'Observations on some Interesting Phænomena in Animal Physiology exhibited by several Species of Planariæ,' Edinburgh, 8vo.

(7) 1815. MONTAGU G.—"Description of several new or rare Animals, principally marine, discovered on the South Coast of Devonshire," 'Trans. Linn. Soc.,' vol. xi, pp. 25-6, pl. v, fig. 3.

(8) 1823. FLEMING, J.—"Gleanings of Natural History, gathered on the Coast of Scotland during a Voyage in 1821," 'Edin. Phil. Journal,' vol. viii, p. 297.

(9) 1826. FABRICIUS, O.—"Fortsaettelse af Nye Zoologiske Bidrag : vi. Nogle lidet bekjendte og tildeels nye Flad-Orme (Planariæ)," 'Kong. Dansk. Vid. Selskab. naturvid Afhandlingar,' Deel ii, Kjöbenhavn, 4to.

(10) 1831. EHRENBERG, C. G.—"Symbolæ physicæ :" 'Phytozoa Turbellaria.'

(11) 1832. JOHNSTON G.—"Illustrations in British Zoology : 3, *Planaria cornuta*," 'Magazine of Natural History,' vol. v, pp. 341-346, fig. 79.

(12) 1836. JOHNSTON G.—"Illustrations in British Zoology : 52, *Planaria subauriculata*," ibid., vol. ix, pp. 16, 17, fig. 2.

(13) 1839. FORBES, E., and GOODSIR, J.—"Notice of Zoological Researches in Orkney and Shetland during June, 1839," Report British Assoc., 9th meeting.

(14) 1840. GRUBE E.—'Actinien, Echinodermen, und Würmer d. adriatischen u. Mittelmeeres,' Königsberg, 4to, pp. 51-6.

(15) 1840. Thompson, W.—" Additions to the Fauna of Ireland," ' Ann. and Mag. of Natural History,' vol. v.

(16) 1844. Oersted, A. S.—" Entwurf einer systematischen Eintheilung und speciellen Beschreibung d. Plattwürmer," Copenhagen, 8vo.

(17) 1845. Kolliker, A.—" Lineola, Chloraima, Polycystis, neue Wurmgattungen, &c.," ' Verhandl. d. schweizerischen naturforsch. Gesellschaft bie ihrer, 29, Versamml. zu Chur.,' Chur. 8vo.

(18) 1845. Quatrefages, A. de—" Études sur les types inférieures de l'embranchement des Annelés—Mémoire sur quelques Planaires marines," ' Ann. Sci. Nat ,' 3me sér., t. iv.

(19) 1845. Thompson, W.—" Contributions to the Fauna of. Ireland, &c.," ' Ann. and Mag. of Nat. Hist.,' vol. xv.

(20) 1845. Johnston, G.—" An Index to British Annelides," ibid., vol. xvi.

(21) 1844-5.—Oersted, A. S.—" Fortegnelse over Dyr, samlede i Christianiafjord ved Droback," ' Kröyers Naturhist. Tidsskrift,' t. i, pp. 415-119.

(22) 1846. Thompson W.—" Additions to the Fauna of Ireland," ' Annals and Mag. Nat. Hist.,' vol. xviii.

(23) 1847. Frey, H., and Leuckart, R.—' Beiträge zur Kenntniss d. wirbellosen Thiere,' Braunschweig, 4to., pp. 82-5, 119-50.

(24) 1848. Schmidt, O.—' Neue Beiträge zur Naturgeschichte d. Würmer,' Jena, 8vo.

(25) 1849. Thompson, W.—" Additions to the Fauna of Ireland," ' Annals and Mag. Nat. Hist.,' 2nd ser., vol. iii.

(26) 1851. Busch, W.—' Beobachtungen ü Anat. u. Entwickel. einiger Seethiere,' Berlin, 4to.

(27) 1851. Schultze, M. S.—' Beiträge zur Naturgeschichte der Turbellarien,' Greifswald, 4to.

(28) 1852. Schmidt, O.—" Neue Rhabdoccelen aus dem nordischen u. adriatischen Meere," ' Sitzungsberichte d math.-natur Klasse d. K. K. Akad. Wiss Wien,' Bd. ix.

(29) 1853. Dalyell, J. G.—' The Powers of the Creator displayed in the Creation,' vol. ii.

(30) 1855. Gosse, P. H.—" On New or Little-known Sea Animals," ' Ann. and Mag. Nat. Hist.,' 2nd ser., vol. xvi, p. 312.

(31) 1857. Harvey, W. H.—' The Seaside Book,' London, 8vo.

(32) 1857. Schmidt, O.—" Zur Kenntniss d. Turbellaria rhabdocœla u. einiger anderer Würmer d. Mittelmeeres," ' Wiener Sitzungsbericht,' Bd. xxiii.

(32a) 1857. Stimpson, W.—" Prodromus descriptionis animalium evertebratorum quæ in Expeditione ad Oceanum Pacificum Septentrionalem, Johanne Rodgers Duce a Republica Federata missa, observavit et descripsit " ' Proc. Acad. Nat. Sci. ' Philadelphia, 1857.

(33) 1860. BENEDEN, J. P. VAN.—"Recherches sur la Faune littorale de Belgique : Turbellariés," 'Mém. de l'Acad. Royale de Belgique.: t. xxxii.

(34) 1861. SCHMIDT, O.—"Turbellarien von Corfu und Cephalonia," 'Zeitsch. f. wiss. Zool.,' Bd. xi, pp. 1–32.

(35) 1861. CLAPARÈDE, ED.—"Recherches anatomiques sur les Annélides, Turbellariés, Opalines, et Gregarines observés dans les Hébrides," 'Mémoires, Soc. Physique et d'Histoire Nat. d. Genève,' t. xvi, pp. 56–80, pls. v–vii.

(36) 1863. CLAPARÈDE, ED.—'Beobachtungen ü. Anat. u. Entwickelungs-geschichte wirbelloser Thiere an der Küste der Normandie angestellt,' Leipzig, fol.

(37) 1865. METSCHNIKOFF, EL.—"Zur Naturgeschichte der Rhabdocoelen," 'Ann. and Mag. Nat. Hist.,' 3rd ser., vol. xvii, p. 57. Translated from 'Archiv f. Naturgesch.,' 31 Jahrg., Bd. i.

(38) 1865. JOHNSTON G.—'A Catalogue of British Non-parasitical Worms in the British Museum,' London.

(39) 1866. LANKESTER, E. RAY.—"Annelida und Turbellaria of Guernsey," 'Annals and Mag. Nat. Hist.,' 3rd ser., vol. xviii, pp. 388–9.

(40) 1868. KEFERSTEIN, W.—"Beiträge zur Anat. u. Entwickl. einiger See-planarien von St. Malo,' 'Abhandl. d. köngl. Gesellch. d. Wissenschft. zu Göttingen,' Bd. xiv, pp. 3–38, pls. i.–iii.

(41) 1870. ULJANIN, W.—"Die Turbellarien der Bucht von Sebastopol" (Russisch), 'Bericht. d. Vereins der Freunde d. Naturwissenschaften zu Moskau,' pp. 1–96, pls. i.–vii.

(42) 1871. LEUCKART, R.—'Bericht über d. Leistungen in der Naturgeschichte d. niederen Thiere,' 1870–1.

(43) 1873. HALLEZ, P.—"Observations sur la Prostomum lineare, Oe.," 'Archives de Zoologie expérimentale,' t. ii, pp. 559–586, planches xx–xxii.

(44) 1874. GRAFF, L.—"Zur Kenntniss d. Turbellarien," 'Zeitschr. f. wiss. Zoologie,' Bd. xxiv, pp. 123–160.

(45) 1875. McINTOSH, W. C.—'The Marine Invertebrates and Fishes of St. Andrews,' Edinburgh, 4to, pp. 105–108.

(46) 1875. MÖBIUS, K.—'Jahresbericht d. Commission zur wissensch. Unter-suchung d. deutschen Meere,' Jahrg. ii und iii, Berlin, fol.

(47) 1878. GRAFF, L.—"Kurze Bericht. ü. fortgesetzte Turbellarien studien," 'Zeitschr. f. wiss. Zool.,' Bd. xxx, suppl., pp. 157–165.

(48) 1878. MERESCHKOWSKY, K. S.—"Ueber einige neue Turbellarien d. weissen Meeres," 'Archiv für Naturgeschichte,' 45 Jahrg., Bd. i. (translated from 'Russischer Arb. d. St. Petersburg Gesellsch. f. Natur-forsch.,' Bd. iv.).

(49) 1878. JENSEN, O. S.—'Turbellaria ad litora Norvegiæ occidentalia (Tur-bellaria ved Norges Vestkyst), Bergen, fol., pp. 1–97, 8 plates.

(50) 1870. HALLEZ P.—'Contributions à l'histoire naturelle des Turbellariés,' Lille, 4to, pp. 1–213, 11 planches.

(51) 1879. LEVINSEN, G. M. R.—"Bidrag til Kundskab om Grœnlands Tur-
bellarie-fauna," 'Vidensk. Meddel. fra. den naturh. Foren. i Kjøben-
havn' (separate copy cited).

(52) 1880. CZERNIAVSKY, V. —"Materialia ad geographiam ponticam com-
paratam," fasc. iii, Vermes, 'Bull. Soc. Imp. d. Natur. d. Moscow,' t. lv.

(53) 1882. GRAFF, LUDWIG v.—'Monographie d. Turbellarien: I. Rhabdocœlida.'

(54) 1884.—LANG, A,—"Die Polycladen," 'Fauna u. Flora d. Golfes v.
Neapel.'

(55) 1886 KOEHLER, R.—"Contributions to the Study of the Littoral Fauna
of the Anglo-Norman Islands," 'Ann. and Mag. Nat. Hist.,' 5th ser.,
vol. xviii (translated from 'Ann. Sci Nat.,' 6me ser., t. xx).

(56) 1891. GRAFF, LUDWIG v.—'Die Organisation d. Turbellaria Acœla,'
Leipzig.

(57) 1891. BÖHMIG, L.—"Untersuchungen ü. rhabdocœle Turbellarien," 'Zeit.
f. wiss. Zool.,' Bd. ii.

DESCRIPTION OF THE FIGURES ON PLATES X, XI, & XII.

Illustrating Mr. F. W. Gamble's paper on "Contributions to a
Knowledge of British Marine Turbellaria."

Pl. X represents the living animals, Pls. XI and XII compression- and other
preparations.

Alphabetical List of Reference Letters for all the Figures.

B. C. Bursa copulatrix. BR. Brain. BS. Spermotheca. CH. Chitinous
portion of copulatory organ. CI. Cilia. COP. Copulatory organ. D. In-
testine. DO. Vitellarium. E. Eye. EI. Ovum. FL. Flagella. GER.
Germarium. GO. External genital aperture. KD. Granule-gland. L. Lens
of eye. M. Mouth. ME. Muscular envelope. MR. Mucous-rods. N.
Nucleus of epidermal cells. OT. Otolith. PA. Packets of rhabdites. PE.
Penis. PH. Pharynx. RH. Rhabdites. RS. Receptaculum seminis. SE.
Spermatozoa. SP. Pharyngeal glands. TE. Testis. V. D. Vasa deferentia.
VG. Vesicula granulorum. VS. Vesicula seminalis. W. Ciliated groove.

PLATE X.

FIG. 1.—Young *Stylostomum variabile*, Lang. Natural length 9 mm. The
three pairs of eyes placed over the brain are very conspicuous, and persist in
the adult (see p. 249). × 65.

FIG. 2.—*Cycloporus papillosus*, var. *lævigatus*, Lang. Natural size. This
form exhibits almost complete similarity in colour, form, and consistency
with the Ascidians on which it is usually found.

FIG. 3.—*Oligocladus sanguinolentus*, Quatref. Length 1·1 cm. × 6.

FIG. 4.—*Cylindrostoma inerme* (Hallez). Length 1 mm. The drawing is made from a specimen slightly compressed. Zeiss obj. C, oc. 1, cam. luc. × 55.

FIG. 5.—*Macrorhynchus Naegelii*, Köll. Length 2·2 mm. This is a colour variety similar to what Claparède observed at St. Vaaste (see p. 206) × 30.

FIG. 6.—*Promesostoma lenticulatum* (O. Sch.). Natural length ·5 mm. The carmine-coloured gut is visible. × 10 ?.

FIG. 7.—*Enterostoma austriacum*, Grff. Natural length ·75 mm. The yellow colour is due to groups of pigment-granules, the black spot to the contents of the intestine. × 40.

FIG 8.—*Provortex rubrobacillus*, n. sp. Natural length ·6 mm. The brown spots are due to the contents of the gut-cells. × 70.

FIG. 9.—*Foria affinis* (Oe.) (probably *Uteriporus vulgaris*, Bergental). Natural length 4·5 mm. The figure is carefully drawn from a freely-moving specimen. The slight lobes of the anterior margin are seen when the animal is viewed from below and in front; this view also shows a slight median projection—similar, in fact, to what occurs in *Convoluta paradoxa*. × 8.

FIG. 10.—*Promesostoma marmoratum* (Schultze). Natural length 1·5 mm. × 20.

PLATE XI.

FIG. 11.—*Plagiostoma dioicum*, Metschff. Natural length ·6 mm. Compression-preparation. × 200.

FIG. 12.—*Provortex rubrobacillus*, n. sp. Natural length ·6 mm. Compression-preparation. × 150.

FIG. 13.—*Promesostoma lenticulatum* (O. Sch.) Compression-preparation. × 150. Concerning the organs marked **R. S.** and **B. C.** see p. 199.

FIG. 14.—*Promesostoma agile* (Lev.). Natural length ·5 mm. Compression-preparation. × 220.

FIG. 15.—Vesicula granulorum and its chitinous investment taken from *Macrorhynchus Naegelii*, Köll. The thickened margin is produced into two curved " spurs." As a rule only one is present. × 220.

FIG. 16.—Copulatory organ of *Promesostoma marmoratum* (Schultze) (partly after v. Graff). × 200.

FIG. 17.—Copulatory organ of *Promesostoma lenticulatum*. × 600.

FIG. 18.—*Automolos* (?) *ophiocephalus* (O. Sch.). The living animal fully extended. × 40.

PLATE XII.

FIG. 19.—Hinder portion of *Cylindrostoma elongatum*, Lev., to show the relations of the genital apparatus. × 500.

FIG. 20.—*Plagiostoma sulphureum*, Grff. Natural length 2 mm. Compression-preparation. × 70.

FIG. 21.—*Automolos horridos*, n. sp. Natural length 1·5 mm. Compression-preparation. × 100.

Plate X.

Fig 2

Fig 1
x 65

BR

Fig 3
x 5

Fig. 5.
x 90

Fig 4
x 55
W
DO
D
GO

Fig 6
x 10

Fig. 7.
x 40
D

Fig 10
x 20

Fig. 8
x 70

Fig. 9.
x 8

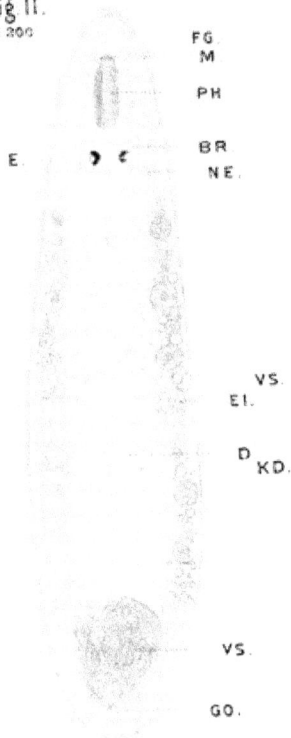

Fig. 11.
× 200

FG.
M
PH
E.
BR
NE.
E I.
VS.
D KD.
VS.
GO.

Plagiostoma dioicum.

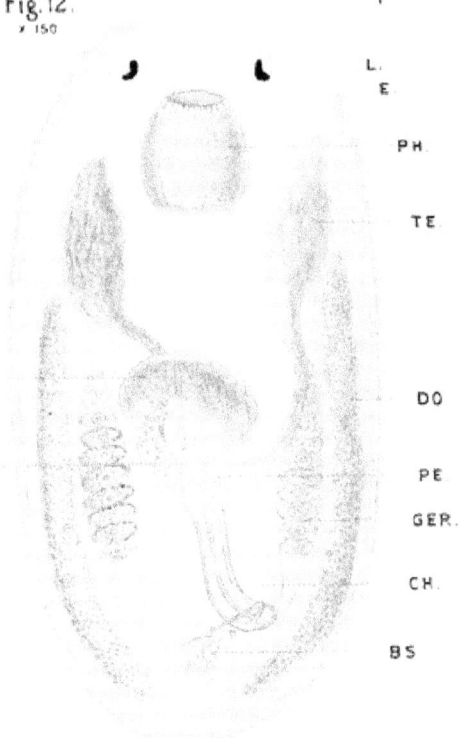

Fig. 12.
× 150

L.
E.
PH.
TE.
DO
PE
GER.
CH.
BS

Provortex rubrobacillus. n. sp.

Fig. 15.
× 220

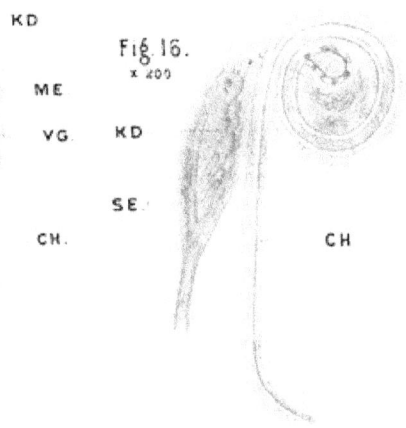

KD
ME
VG. KD
SE.
CH.

Fig. 16.
× 400

CH

E

L

DO

DO

VS

PH

GER

RS

TE

COP.

K.D.

GER

Promesostoma lenticulatum

Promesostoma agile

VS

OT

Fig. 17.

Fig 18

KD

Automolos ?ophiocephalus

Fig. 19
× 500

PH
EI

D
DO
GER.
PE
N
KD
MR
GO

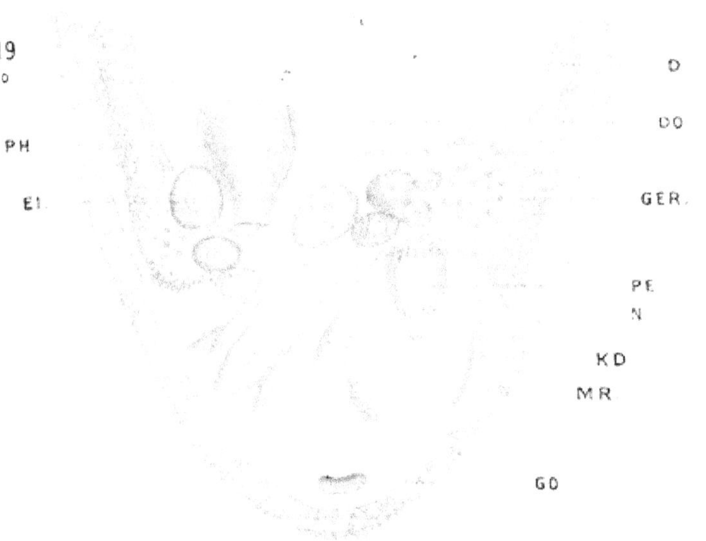

Cylindrostoma elongatum, Lev.

Fig. 20.
× 70

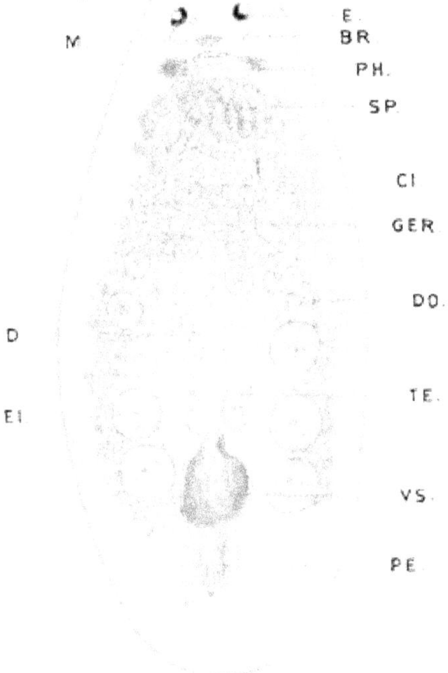

M

E.
BR
PH.
SP.
CI
GER
DO.
TE.
VS.
PE.

D
EI

Plagiostoma sulphureum n.sp.

Fig. 21.
× 100

OT
BR
D.
GER.
PH
DO.
RH.
PE.

FL
TE
CI.
VD.
VS
PE.
PA

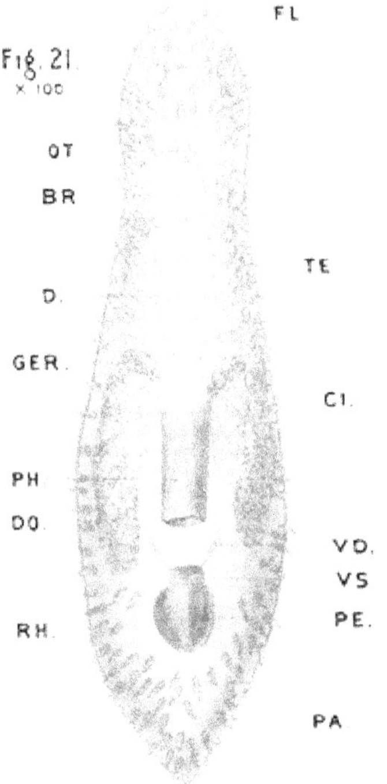

Automolos horridus, n.sp.

From the ANNALS AND MAGAZINE OF NATURAL HISTORY,
Ser. 6. Vol. xii., *September 1893.*

On an Abnormal Specimen of ANTEDON ROSACEA. *By* HERBERT
C. CHADWICK.

Plate XIII.

Three months ago, while selecting specimens of *Antedon rosacea* for
serial section-cutting from a number which had been forwarded to the
Zoological Laboratory of the Owens College by the authorities of the
Zoological Station at Naples, my attention was arrested by one to the
disk of which a small rounded body was attached. A cursory examination at once showed the specimen to be one of very exceptional interest, and my thanks are due to Prof. Milnes Marshall for permission
to examine and describe it.

The disk (Pl. XIII. figs. 1 and 2), which measured 7·5 millim. in
diameter, bore the usual number of well-developed arms, and with the
exception of the displacement of one of the ambulacral grooves, to be
more fully described later on, was in all respects quite normal. On
its oro-lateral border, however, it bore the body to which allusion has
already been made, and which proved to be a supernumerary disk
(figs. 1, 2, and 2, *s.d.*). Roughly spherical in shape and about 3
millim. in diameter, it was attached to the normal disk by a sort of
stalk, which gradually narrowed from the oral to the aboral surface.
Near the centre of its oral surface was a well-developed mouth, fringed
with tentacles, from which five ambulacral grooves radiated, just as
do those of the disk of a normal *Antedon.* Of these, four could with
little difficulty be traced outwards to the aboral aspect.

The remaining one (figs. 1 and 3, *x*) ran along the stalk of attachment to the normal disk and joined the ambulacral grooves of the
pair of arms nearest to it, immediately after crossing the line of junc-

tion of the two disks. On the aboral surface the anus appeared as a minute crescent-shaped aperture (figs. 2 and 4, *a*). Close to it was a minute, scarcely-distinguishable pore, another rather larger aperture appearing on the summit of the funnel-shaped projection, *f.p.* (figs. 2 and 5). The nature and connexion of these will appear later on.

Minute Anatomy.—Having carefully noted and drawn the external characters of the specimen, I decalcified it by immersion for twenty-four hours in a 10 per cent. solution of nitric acid, and, after staining in borax carmine, I was fortunate enough to obtain an unbroken series of sections by means of the rocking microtome. From a very careful study of these I find that the body-cavities of the two disks communicate freely with each other through the stalk or isthmus of tissue which unites them, their alimentary canals, on the other hand, being quite distinct. The alimentary canal of the supernumerary disk (figs. 3 and 4, *g*) is well developed and contains food. The ambulacral system is also well marked and presents a feature of special interest. The minute pore close to the anus, to which I have already alluded, opens into a canal-like space (fig. 4, *c*), which traverses the body-wall for a distance equal to the thickness of seventeen sections, and again communicates with the exterior through the funnel-shaped projection already described (figs. 2 and 5, *f.p.*). That this canal was a modified ambulacral groove is shown by the epithelial cells which line it. They are precisely similar to those which line the ordinary ambulacral grooves; and further evidence in the same direction is afforded by the presence in its walls of numbers of the deeply staining problematical bodies which are invariably seen in sections of the ambulacral grooves of this species. Beneath the epithelium of the ambulacral grooves the nerve-band can be recognised without difficulty in most sections. The circular water-vessel (fig. 5, *c.w.v.*) and radial water-vessels are also present, and from the former a considerable number of water-tubes (fig. 5, *w.t.*) depend into the body-cavity. Water-pores traverse the body-wall in all the sections and are abundant on the interambulacral area, marked with an asterisk in fig. 1 (see also fig. 3, *w.p.*). The skeletal and axial nervous systems present in the normal disk are entirely absent in the supernumerary one; so also is the central plexus.

The interesting question now arises—What was the mode of origin of the supernumerary disk ?   In answer to it two hypotheses, may, I think, be advanced :—

1. That the supernumerary disk originated as a bud from the normal disk.

2. That it is the result of incomplete evisceration.

In favour of the former hypothesis is the intercommunication of the body-cavities of the two disks—a condition of things one would expect to find in a budding organism.   Against it is the entire absence of arms, skeleton, and axial nervous system in the supernumerary disk. The comparatively large size attained by the supernumerary disk and the fact that the remaining systems of organs had attained their adult condition add importance to this objection.   A still weightier objection lies in the fact that, so far as I know, the formation of a bud has never been observed in any Echinoderm.

I am indebted to Prof. Marshall for the second hypothesis, and it appears to me to explain the facts most conclusively.

Though *Antedon rosacea* has never been proved to eviscerate spontaneously, eviscerated specimens frequently occur in dredgings ; and the experiments of Prof. Marshall* and Mr. Dendy† have shown that evisceration may be and often is followed by complete regeneration of the visceral mass.

These facts seem to me to make more than probable the supposition that at an earlier period the specimen had suffered evisceration without the visceral mass being completely detached.   By the continuity of the ambulacral grooves of two of the arms of the normal disk with one of the grooves of the supernumerary disk a supply of food would be ensured to the latter without seriously curtailing that of the former during regeneration.   In the paper just cited Mr. Dendy has shown in how short a time the visceral mass may be regenerated, twenty-one days being a sufficient length of time for regeneration to become so complete that "there is little to distinguish a regenerated specimen of this date from a normal *Antedon* except the small size of the visceral mass and the want of pigment upon it."

* "On the Nervous System of *Antedon rosacea*," Quart. Journ. Micr. Sci. xxiv. (1884) pp. 507-548.

† "On the Regeneration of the Visceral Mass of *Antedon rosacea*," Studies from the Biological Laboratories of the Owens College, i. (1886) pp. 299-312.

The abnormal character and displacement of the anus and the canal-like ambulacrum are not so easily accounted for; but they are minor points, and do not appear to me to impair the value of what has been advanced above.

EXPLANATION OF PLATE XIII.

List of reference letters.

a. Anus.	*g'.* Gut of supernumerary disk.
a.g. Ambulacral grooves.	*m.* Mouth.
e. Abnormal ambulacrum.	*r.w.v.* Radial water-vessel.
c.w.v. Circum-oral water-vessel.	*s.d.* Supernumerary disk.
f.p. Funnel-shaped projection of	*s.o.* Skeletal ossicles.
supernumerary disk.	*w.t.* Water-tubes.
g. Gut.	*x.* Ambulacral groove.

Fig. 1. Oral surface of abnormal specimen of *Antedon rosacea*, × 5.

Fig. 2. Aboral surface of abnormal specimen of *Antedon rosacea*, × 5.

Fig. 3. Sagittal section through the normal and supernumerary disks, showing the point of union of the two, × 16.

Fig. 4. Sagittal section of the supernumerary disk, passing through the mouth and anus, × 16.

Fig. 5. Sagittal section of the supernumerary disk, showing the funnel-shaped projection traversed by the abnormal ambulacrum, × 16.

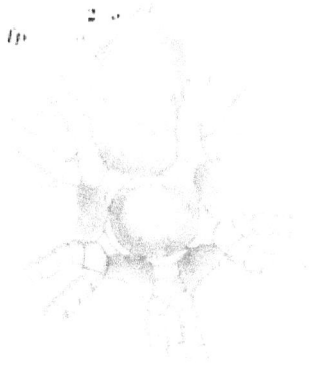

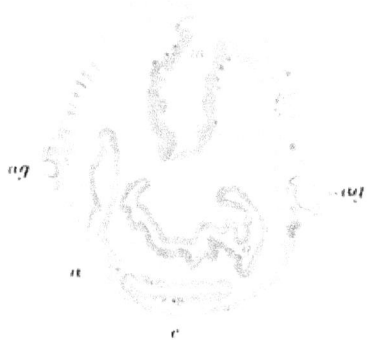

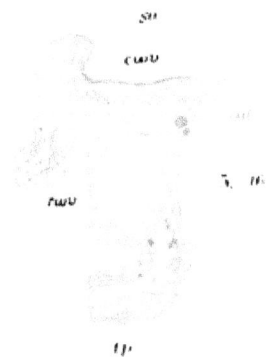

THE STRUCTURE AND HABITS OF ARCHÆOPTERYX.

By C. H. HURST, Ph.D.

Plates XIV., XV., XVI.

I.—THE SKELETON OF ARCHÆOPTERYX.

Apart from a single feather, only two specimens of *Archæopteryx* are known, and it is possible that these may not be identical in species or even in genus. So far as we know them, the differences between the two appear, to those who are best qualified to judge, to be too small to justify separation into two species. Though both were found in Bavaria, I shall refer to them as the "Berlin specimen" and the "London specimen" respectively.

It is not convenient to begin with a description of the external form of the bird, as is customary with recent species, for that external form can only be guessed at with reasonable chance of guessing accurately after a careful consideration of the structure of such parts as are still preserved. This is even more conspicuously true of the habits of the animal.

Of the skeleton, if we assume the two specimens to be so nearly related that the characters exhibited in either may be taken as true of both, we have quite an extensive knowledge.

The *vertebral column* is readily divisible into four regions: cervical, trunk, sacral, and caudal. Whether the vertebræ are fully ossified or not it is difficult to say. I can find no justification for the statement that they are amphicoelous. Professor Dames tells me that his statement to that effect is a mere slip of the pen, and that he intended only to say that, so far as can be seen in a specimen in which the vertebræ are still in their natural relations with one another, the ends are flat and not as in most birds, saddle-shaped. The central or internal part of each vertebra in the London specimen is stated by Owen to be represented by a deposit of crystalline "sparry matter" in the caudal region, while the outer "crust" has adhered to the upper slab or "counterpart." Whether this really shows that the vertebræ (of the tail) were mainly cartilage or other soft tissue with only a crust of bone or not, may be open to question. The perfectly-fitting joints, the large transverse processes of the anterior caudal vertebra, and the slenderness and stiffness—as shown by the straightness of the tail in both specimens—of this region of the vertebral column are strong evidence that the bones were *well-ossified*.

Of the nine *cervical vertebræ*, only eight are well-preserved, the first being almost unrecognisable. Measuring the lengths of the centra of these on a large photograph (scale ¹³⁷⁄₅₇), I make the sum of the eight in the Berlin specimen to be about 75 mm.; but Professor Dames gives numbers which together make only 60·5. A glance at Plate XV. will show the position of the neck in this specimen. It is very strongly arched so as to bring the head almost into contact with the back of the animal in the region of the thorax. It is difficult to make these measurements accurately in either the specimen or the photograph, but the discrepancy between the two measurements is too great to be accounted for by this difficulty, and I suspect that Professor Dames' measurements have been made along the inner curve—*i.e.*, through the neural spines—while mine were made near the ventral curve, *i.e.*, through the centra of the vertebræ. I suspect, therefore, that when the animal's neck was straightened out it would be 75 mm. long in addition to the length of the atlas, which may be taken to be a very small quantity as in modern birds. Of the nine cervical vertebræ the middle ones are longer than those nearer the ends of the neck, the fifth being the longest.

Cervical ribs, apparently movably articulated, may be made out, and there appear to be eight pairs of them. The neural arches and spines are well-developed and strong, the spines being 2 to 3 mm. high.

The *trunk vertebræ* being somewhat displaced, and the vertebral column distorted, it is not very easy to make sure of their number. There appear, however, to be ten, measuring together about 70 mm. The vertebræ appear to be almost equal in size, and nine of them bear *ribs*. There are also *ventral ribs*, resembling the "*abdominal*" ribs of the geckos and chamæleons, and clearly showing the ventral boundary of the abdominal cavity (*see* 14 in Plate XV.).

The *sacrum* is hidden in the Berlin specimen except at its ends.

It measures 26 mm. in length. It is probable that there are about seven sacral vertebræ.

The *vertebræ of the tail*, twenty in number, measure together about 170 mm.—slightly less perhaps. The first few are very short and stout, each measuring about 4 mm. in length and 4 mm. in height. The first four have well developed transverse processes; in the fifth this process is not well preserved, and the vertebræ behind this have no transverse processes, but only a ridge. The vertebræ are longest nearer the middle of the tail, the eleventh measuring nearly 12 mm. The tail as a whole seems to have had little flexibility, for it is almost perfectly straight in both specimens. The tale of the London specimen has apparently only eighteen vertebræ and measures 180 mm.

The *skull* has been much further exposed since the photograph was taken. It is large and fairly massive, the jaws are stout, and teeth are very easily made out in the upper jaw. Those of the lower jaw are, however, hidden by those of the upper, and it is impossible to say at present how many there were. The sclerotics are ossified. The hinder part of the skull is destroyed in the Berlin specimen, and it is worthy of note that the cranial cavity was not filled with matrix. No part of the skull is recognisable with certainty in the London specimen, though it may be that the supposed cast of the brain (!) is a portion of the skull.

The *ribs*, both vertebral and ventral, are very slender. There are no uncinate processes visible.

Of the *sternum* nothing is known, though much has been written. In the Berlin specimen it probably lies still hidden in the matrix. The position of the ventral ribs shows that it must have been small.

The *scapulæ* in the Berlin specimen were broken in exposing the specimen. The right one is easily recognisable in Plate XV. They are flat curved bones, not unlike those of a modern bird. Their length is 43 mm. or thereabouts, according to Dames. In the photograph only a portion is seen

The *coracoids* are in the Berlin specimen largely hidden. I have not specially examined what portion is exposed in the London specimen. The dorsal ends are exposed in the Berlin specimen and possess a furcular tuberosity as in other birds.

Of the *furcula*, a small portion is seen at the left shoulder of the
Berlin specimen. It was, however, imperfectly exposed at the time
when the photograph was taken. A larger portion is seen in the
London specimen. It is a characteristically avian furcula, U-shaped
ventrally, and articulating with the furcular tuberosity of the coracoid
at each shoulder.

The *humerus* is a well-developed bone in each wing. Its form and
dimensions may be seen in Plates XIV. and XV. It differs from that
of other birds in being devoid of the pectoral crest or ridge for the
insertion of the great pectoral muscle. As Dames points out, this
confirms his view that the sternum must have been small, as must
also the great pectoral muscle. In Plate XIV. the proximal end of
the humerus is covered by a portion of the matrix, which has since
been removed (at 11 in Plate), and that plate consequently gives an
impression of a humerus which is slightly shorter than the true length.

The bones of the fore-arm, seen in Plates XIV and XV, are a
straight *radius* 55 mm. long, and a *curved ulna* 56 mm. long.

The *carpus* offers great difficulties. Owen figures two bones, one of
which is visible in the London specimen. Why he should ignore the
enormous ulnar carpal, which is a conspicuous object in the London
specimen, need not here be discussed. It is conspicuously shown in
Fig. 2 and in Plate I of Owen's memoir, where it is numbered 56' and
described (presumably with the radial carpal) as "left carpus" (it
being probably a part of the right carpus), and something wholly unlike
it is put in its place, in dotted lines, in his second plate (Fig. 2).

Of these bones I have seen two clearly, one being the radiale (4 in
Plate XIV), which is visible in both the specimens, the other the
"ulnare," visible only in the London specimen. In the Berlin specimen the carpus lies radial side uppermost, and it is not surprising
that, like some other parts, the ulnar portion of the carpus lies still
embedded in the matrix. This is even admitted by Dames. The
little bone called "ulnare," and drawn from imagination by Owen, and
also drawn by Dames, may or may not be present. I have tried, and
failed, to make it out in the Berlin specimen, and I have also tried,
and failed, to make sure that it is not there. One thing only I can
say of it, viz., if present it is probably the *intermedium*, and not the
ulnare. The "ulnare" is the enormous and conspicuous bone shown

at the distal end of the right radius and ulna. It is for a carpal bone, of enormous size, and I am not prepared to believe that it played no part in the support of the metacarpals.

Of the distal row of carpals it is only possible to say that they are not recognised in either specimen. Whether they have fused with the metacarpals, as they do in modern birds, or were cartilaginous and so not preserved, or were fused with the bones I have referred to as belonging to the proximal row; or whether the two figured by Owen and Dames are the proximal row, and the large bone I have called "ulnare" is really, as the London specimen suggests, a fused mass representing the whole or part of the distal row of carpals, can only be decided, so far as I can see, by one of two consummations "devoutly to be wished"—(1) the excavation of the exceedingly thin and fragile Berlin slab *from the back*, or (2) the discovery of fresh specimens. The first of these involves too great a risk to what it is hardly an exaggeration to say is the most valuable palæontological specimen in any museum in the world.

To admit that one does not know what that bone is, is one thing; to ignore its existence is another. Whether it be right or wrong, I shall for the present call it the *ulnare*. Subsequent proof that it is something else, *e.g.*, a crocodilian "lenticulare," or, as seems not improbable, *unciforme*, will not invalidate my argument.

The *hand* has been much misrepresented both in words and in drawings. There are *five digits* and no fewer, and I never suspected that it would be necessary for me to give further proof than that already given in my essay on errors. This conclusion, however, having been controverted, I will venture now to prove it over again by three distinct proofs, each of which is in itself conclusive.

(1.) Three *long, slender* fingers on each hand are plainly seen on the Berlin slab. They are made up of two, three, and four phalanges respectively, in addition to a metacarpal each. Each bears a claw, which, though not easily made out in the photographs, especially in the smaller photographs, is perfectly distinct in most cases in the slab itself. There can be no doubt, and nobody does doubt, that these three correspond to the digits I, II, and III respectively of the normal pentadactyle reptilian fore-limb. The lengths of the various metacarpals and phalanges in the Berlin specimen are as follows, beginning at the proximal end, *i.e.*, with the metacarpal, in each case :—

I. $8 + 20 + 11 = 39$ mm.

II. $27.5 + 15 + 19 + 13 = 74.5$ mm.

III. $26 + 5.5 + 4 + ? + ? = 44.5$ mm. The joint between the third and the ungual phalanx is hidden, but these two together measure 19 mm. Of these bones the second metacarpal is the largest, and at its basal end it is under 4 mm. thick.

Some of the bones corresponding to these are to be seen in the London specimen: but as they are displaced, it is not possible to identify them with certainty. What Owen called the two terminal phalanges of the digit I, closely resemble the two terminal phalanges of digit II of the Berlin specimen, and I take them for these phalanges. They measure respectively 22 mm. and 15 mm., i.e., they exceed the bones of the Berlin specimen in the proportion of rather over 9 to 8. To justify this determination I give the lengths of some other bones in the specimens. The first number in each case is the length of the bone in the Berlin specimen; the second, that in the London specimen. Ulna, 56 mm., 63.5 mm.; Radius, 55 mm., 62 mm.; Femur, 51 mm., 58 mm. (?); Tibia, 71 mm., 81 mm. In each case except that of the femur the ratio is almost exactly 8 : 9, and in the case of the femur it is impossible to measure the exact length in the Berlin specimen. The numbers of vertebræ in the tails differ in the two specimens, so that it will not be safe to take the ratio in length of the two tails as a guide. There is no other bone which can be identified and measured with certainty in both specimens, so we may adopt 8 : 9 as the relative sizes of the Berlin and the London specimens respectively.

But in thickness a different relation holds. In corresponding bones of two similar animals we find that the ratio of thickness to length is always greater in the larger animal. And this is true here: all bones of the London specimen are stouter and more massive than those of the Berlin specimen. Now, in the London specimen two conspicuous bones were identified by Owen as the "third" and "fourth" metacarpals. They measure 39 and 33.5 mm. respectively in length. In thickness they are much greater than any hand-bone of the Berlin specimen. Others have regarded these bones as the second and third metacarpals Suppose this were the case, then we get these ratios between the London and Berlin specimens.

39 : 27·5 = more than 11 : 8 and 33·5 : 26 which is 10·3 : 8. Further, the bones are utterly unlike their supposed equivalents in the Berlin slab. They are far stouter, and the longer of the two is exceedingly broad at the base, and therein is well-fitted to resist torsional stress or twist at the joint. In their proximal halves, instead of being slender and almost circular in section, they are stout and have ridges which, when the two were fitted together, would have prevented their movement one on the other. Whatever they are, they are utterly unlike any bones visible in the Berlin specimen. Their position with reference to the feathers of the wing, in spite of the dislocation of other bones, is just that of the large metacarpals in an ordinary bird's wing; and the fact that these feathers are still in their normal position in this wing (the left) justifies the belief that when the animal finally settled down previous to fossilisation those feathers were still bound to those metacarpals by ligament.

This is proof no. 1 that those two bones are the metacarpals of the digits IV and V.

(2.) The second proof is a more formidable one. Some hundreds of experiments extending over hundreds and even thousands of years have shown the effect of "selection" upon dogs, horses, sheep, pigs, pigeons, poultry, vines, roses, plums, apples, pears, strawberries, gooseberries, blackberries, pansies, daisies, dahlias, chrysanthemums, etc., etc., and the result is the same in all cases. Selection occurs in Nature (Naudin, Darwin, Wallace), and its effect is the same as in the case of artificial selection (Naudin, Darwin, Wallace, Bates, and others). I do not think it necessary to repeat the proof of this statement here : the proof is far too long, too well known, and too widely accepted for me to need to say more about it.

We may, therefore, take it as proved that the form and dimensions and structure of every bone and feather in *Archæopteryx* is the outcome of long-continued Natural Selection. The form and structure of the bones of the three digits visible in the Berlin specimen, and of the feathers in the same specimen, show what the conditions of selection have been, and what have been the uses of those several parts.

The digits I, II, and III are long, *slender*, and clawed. Each metacarpal and phalanx is concave on the flexor surface. The ends of the bones are curved like pulleys, allowing of free movement at every

s

joint. A distinct tubercle, for the insertion of the flexor tendon, is recognisable at the proximal end of almost every one of the eighteen phalanges, and these, together with the curvature of the bones, show that flexor muscles were well developed and active and useful, and, in view of the forms of the joints, were useful in producing extensive flexion of those digits. One of these joints has been referred to by some who had not seen the specimen as possibly a fracture (the joint between the second and first phalanges of the third digit, marked 16 in Plate XV); but a more perfect joint does not exist in the toe of any existing bird than that joint. It is perfectly preserved, and nobody who has seen it can doubt for a moment that it has been evolved under the influence of Natural Selection, and was exceeding well-adapted to allow of a very extensive flexion.

The feathers are as perfectly adapted to resisting the passage of air through them as in any modern bird. Those who have studied the mechanism of flight in detail will recognise why those feathers are all so curved that the dorsal surface of the wing when at rest is *convex*; why the anterior division of each vane is narrower than the posterior one, and is strongly curved and overlaps the posterior division of the vane of the next feather in front. They will know that such a wing is useful only if adequately supported by rigid bones capable of resisting very considerable torsional stress.

These two sets of structures—the digits I, II, and III, and the feathers—have been evolved under the direction of Natural Selection. They are both, therefore, fitted to perform the functions they actually did perform. The digits were useful for some purpose involving extensive flexion: they were, in fact, used to grasp parts of trees—for I shall show later that the animal did not habitually walk on the ground on all fours. They could not do this if those large feathers were attached to them. Mr. Pycraft has shown that in *Opisthocomus* the young use the digits for climbing, and that in order to enable them to do so the development of the mid-digital and ad-digital quills is delayed till such time as the young are able to fly or to climb without the help of these digits. I thank him for this excellent illustration (NAT. SCI., vol. v., pp. 355 and 358). It confirms the opinion I have expressed that flexible digits cannot be used for climbing if they bear large quills.

And again, the digits I, II, III of *Archæopteryx* (which the large size and perfectly ossified bones show to be an adult, as also do the well-developed feathers) are, by virtue of their very great slenderness and narrowness at the joints, incapable of resisting a great torsional stress.   Unless those feathers exert a great torsional stress on the bones supporting them they are useless.   I have shown they were not useless.   Therefore they exerted a great torsional stress, and therefore they were supported by bones not yet seen in the Berlin specimen, although those of the left wing are seen in the London specimen.   It follows, therefore, that the first three digits were used for climbing, and that one or more others were present to support the feathers.

(3.) The third proof is incomplete.   It shows only that the digits I, II, and III did not support the feathers, and that, therefore, something else must have existed to do so.   Its simplicity is unsurpassed. It will appeal even to those who ignore both the principles of mechanics and the action of Natural Selection.   The figure 10 is placed on the surface of the right wing in Plate XV.   In this region the dorsal surface of the wing is convex.   A rule or "straight-edge" placed on the wing across this point, parallel to the ulna and resting upon the first and second digits, touches the wing along the whole of its length from number 10 backwards.   In front of this the feather-surface curves downwards, so as to be perhaps 2 mm. below the edge of the rule near the digits.   The *lower* surface of the metacarpal and of the first and second phalanges of the *second* digit lies fully 1 mm. above that feather-clad surface.   The bones of the *third* digit are closely pressed down upon, but not sunk below, that surface.   Therefore those digits did not lie *in* but *upon* the feathered wing when that animal finally sank dead upon the mud in which it has been preserved.   Therefore, further, other bones (or bone) were present to support those feathers. No argument from embryology will shake that conclusion.

The *pelvis* is seen in the London specimen only, and in this specimen nothing is to be learnt from the left innominate, while even the right one is imperfect.   This innominate appears to have been about 50 mm. long.   The acetabulum is perforate.   I believe there is no anti-trochanter, though in absence of the specimen I would not make the statement definite.   It is characteristically avian pelvis so far as concerns the length of the ilium and its prolongation to about an equal extent behind and in front of the acetabulum. If I mistake not, it is conspicuously unlike the pelvis of any existing bird in the matter of width, and the bearing of this will be shown in the sequel.

The *femur*, more slender than in existing birds of the same size, is strongly curved, the flexor surface being concave.

The *tibio-tarsus* is almost perfectly straight, and has only a small cnemial crest. The *fibula* is complete but slender.

The *foot* is a characteristically avian foot. In the Berlin specimen the matrix around the feet is so hard that a complete exposure of them has proved impracticable. The London specimen shows the left foot well. It is more massive and in every way larger than the corresponding parts of the Berlin specimen.

II.—THE FEATHERS.

The evidence before us does not justify the expression of any opinion as to whether the body of *Archæopteryx* was covered with feathers all over or only partially. Three chief kinds of feathers are, however, recognisable in the fossils :—(1) quills, (2) coverts, and (3) contour feathers.

The *quills* are exceedingly well-preserved, especially in the Berlin specimen. The *remiges*, or wing-quills, had the characters of those of many ordinary birds, such as a pigeon. The calamus is not clearly seen, as it is hidden by the coverts; but the narrowing of the vane near the base shows (*e.g.*, in the second and third primary quills of the left wing of the Berlin specimen) that its length was much the same in proportion to the rest of the feather as in the corresponding feathers of a pigeon. The rachis is clearly seen, and is slightly curved so as to render the ventral surface of the wing concave and the dorsal surface convex. It tapers gradually as in feathers of the usual type. The groove seen along the dorsal surface of the rachis is probably due to shrinkage of the medullary substance during fossilisation—but this point is open to dispute. The vane, as in nearly all birds, is curved, the anterior and narrower moiety much more strongly than the posterior, which latter is overlapped dorsally by the anterior portion of the feather next following it. The barbs are easily seen, even in photographs; but I have been unable to make out the barbules

with certainty. Their existence, and even that of the hooks which
serve to maintain the relative positions of the barbs, is safely to be
inferred from the perfect regularity with which the barbs lie side by
side in the Berlin specimen.

Of the quills, there are in each wing seven primary and ten
secondary. The lengths of the primary quills, *i.e.*, of the quills borne
by the metacarpals and phalanges, are as follows (commencing with
the first) :—65, 90, 120, 125, 135, 130, and 120 mm. The secondary
quills, borne by the ulna, are not easy to measure accurately, but they
diminish gradually from the carpal region to the elbow. The first is
115 mm. long ; the last or tenth, 75 mm. Taken as a whole, these
remiges, though less numerous than in most modern birds, are as
perfectly fitted, by their form and arrangement, for the purpose of
flight as in, say, a pigeon. Their size, though not difficult to determine absolutely, is difficult, if not impossible to determine relatively
to the weight of the body ; for in our guesses at the weight of the
animal a very large margin must be left for possible error.

The *rectrices*, or tail quills, differ very remarkably from those of any
other known bird. Unfortunately, I overlooked the question as to
their number when I was in Berlin. Previous authorities regard
them as arranged in pairs, one pair to each vertebra of the tail, but
Dames is very cautious on this point. My large photographs suggest
that they are somewhat more numerous, but the point is one which
may be left for the present undetermined. They differ from the corresponding feathers of modern birds of flight chiefly in size, being very
much smaller than those of ordinary birds of equal size. Those of the
anterior part of the tail are about 50 mm. long, or perhaps a little less
(Berlin specimen). Further back they are longer, the maximum
length of about 95 mm. being at about the twelfth caudal vertebra.
To what extent, if any, they could be spread out and closed together
we can only guess, and the fact that they lie in both specimens at an
angle of about 30° to the axis of the tail does not help us much; for
if the animal had the power of spreading its tail feathers, this would
perhaps have been effected by means of a muscle arising from some
bone in the pelvic region, through the mediation of a slender tendinous
band running along each side of the vertebral column of the tail, and
opposed by a slender elastic ligament which would, when the muscle

was relaxed, bring the feathers into some such position as that which they occupy in the fossils known to us. These tendons and ligaments may well have been very small, and the absence of any trace of them in the fossils is not sufficient justification for our stating that they did not exist. We must, therefore, remain in ignorance as to whether *Archæopteryx* could or could not spread and close its tail.

Though the feathers are small, their great number gives to the tail, looked upon as an aëroplane, a very considerable surface, and this surface is greatly increased by the development of a series of feathers, which may for convenience be classed as rectrices, along the sides of the hinder part of the trunk. So far as I can make out by means of drawings that I have made to scale of the bird in the flying position, these lateral rows of feathers constitute with the tail-feathers a continuous aëroplane, extending forwards as far as the posterior edge of the extended wings. All those who have made drawings of the animal "restored," so far as I know, ignore the existence of this lateral aëroplane, and represent the lateral feathers of the tail-series as coming to an end in the tail, and the London specimen certainly suggests that this is the truth of the matter : the Berlin specimen, however, leaves no room for doubt as to the accuracy of the account I have now given.

Archæopteryx, unlike any other known bird, bore quills on its tibiæ. These are not either remiges or rectrices—and, indeed, the lateral aëroplane in front of the tail is not strictly made up of rectrices. In the absence of a better name, I will call them *tibial quills*. These *appear* to have lain in a single plane, which is the plane of flexion of the leg, the plane in which femur, tibia, and metatarsals all lie when the limb is bent; and they were apparently arranged in two series along the surfaces corresponding to the anterior and posterior surfaces of the human tibia, *i.e.*, the extensor and flexor surfaces. The number of them cannot be made out with certainty. The longest appear to have measured a little over 30 mm. in length. How they were placed in reference to the muscles I cannot say, and though they appear to have lain in a single plane, they may perhaps—though I do not believe it— have been "breeches," as they have been described. They extended along the whole length of the tibia, and certain appearances in the region between the left leg and the tail in the Berlin specimen suggest that the flexor-series extended also to the region of the femur.

Coverts are recognisable with certainty over the primary and secondary quills of the wings, especially the left wing, but they are not nearly so well preserved as the more robust quills. They appear to have been very slender, and now lie at an angle of about 40° to the quills. This deflection may have been due to the action of a stream of water passing over the dead bird when it first came to rest where it was finally fossilised; and, if so, the animal must have come to rest with its head up-stream, as one would naturally suppose.

The appearances in the fossil do not justify any statement as to the coverts in the tail or over the tibial quills.

Contour feathers may be recognised with certainty only in the cervical region. Three are well preserved between the right hand and the label bearing the number 11 in Plate XV. I believe I have recognised others ventral to the fourth and fifth cervical vertebræ; but this is far from certain. As to the covering of the rest of the body, we are in the dark.

III.—HABITS.

Archæopteryx was an arboreal quadruped fitted for flight, if not for prolonged flight.

First, as to quadrupedalism and attitude.—I have already (p. 272) pointed out that the digits I, II, III of the hand are long and slender and flexible; and that each metacarpal and phalanx of these digits is curved, the concavity being ventral; and that the tubercles for the insertion of the flexor and extensor muscles of some of these phalanges are distinctly recognisable. To this I would now add that their joints are at different levels, showing that each finger could be flexed independently of the rest, and therefore that they were *free* and not bound together. I also showed that they did not support the quills, but were free from the wing except at their carpal ends. If they were bound together in the wing they would be inflexible, inasmuch as the joints of each digit are at different levels from those of the other digits.

The absence of that shifting backwards of the heavy abdominal viscera which is seen in such bipeds as birds, squirrels, kangaroos, dinosaurs, etc., and the lizard-like form of the body, are so clearly shown by the ventral ribs that it is obvious that its centre of gravity would be in front, not only of the acetabulum, but of the knee. If

the animal walked, or even stood, on two feet at all, it would have had
to stand either bolt upright or with its dorsal surface directed slightly
downwards. Whether the tail would then be on the ground (as in a
kangaroo) or in the air (as in a squirrel) is open to doubt, for we do
not know enough about the flexibility of its proximal portion to be
able to say with certainty whether it could or could not be bent up
over the animal's back. The presence of a long, heavy neck and
massive head, which were supported by an elastic ligament, as shown
by the curvature of the neck in the fossil (as in *Compsognathus* and
the pterodactyles), add greatly to the force of this argument. A duck
has to walk with its body tipped up almost on end, in spite of the
great shifting of the heavy organs of the abdomen backwards between
the legs. There is no room for doubt that in *Archæopteryx* the centre
of gravity would be much further forward, not only on account of the
heavy head and neck, but also on account of the solidity of the wing-
bones, as shown by the absence of pneumatic foramina. Let any who
doubt the justice of this argument compare the pelvis as seen in
Plate XVI (which the authorities of the British Museum have kindly
allowed to be prepared in illustration of this article) with the pelvis of,
say, a pigeon or a dinosaur, or any other animal whatever capable of
walking on two limbs in any but an erect position. The small size of
the cnemial crests, moreover, forbids us to believe that the hind-limbs
alone were able to bear the weight of the body when the knee was
bent.

This bird was not only not a biped, but it did not walk on the
ground at all. It would have been as helpless on the ground as a bat
or even a sloth. The great length of the hind-limbs and shortness of
the fore-limbs, at any rate when these latter were so flexed as to keep
the feathers off the ground ; the position of the shoulder joint, and
especially of the articular surface of the humerus : these render the
animal unfit for such a habit as even quadrupedal locomotion on the
ground. The perfect state of the wing-quills at their tips shows that
they were not brought habitually into contact with the ground, and I
know of no rational argument in favour of the view that the animal
did live largely upon the ground. For quadrupedal locomotion in
trees the animal is admirably adapted. Long, flexible digits, provided
with claws on all four limbs, fit it at least as perfectly to arboreal

quadrupedalism as *Galeopithecus* or *Pteromys* or any other "flying"
mammal; and I think, from the greater length and flexibility of those
digits, fit it even more perfectly for such a habit than even these
mammals, and incomparably more so than the bats with their
backwardly-directed hind limbs, which, while they serve to enable the
animal to hold on to the tree or other body, are so modified for the
support of the wing as to be of little use for quadrupedal locomotion.

As to flight.—*Archæopteryx*, though less well fitted for prolonged
flight than most modern birds, was certainly capable of flight. As
some have maintained that it was well fitted for powerful and prolonged
flight, I will mention the features pointing to an opposite conclusion.
The absence of a pectoral crest on the humerus, the small size of the
sternum (*see* p. 269), and consequent relatively small size of the
great pectoral muscle, indicate a deficiency of propulsive and sustaining
power in flight. The absence of pneumatic foramina and consequent
absence of air-cavities in the wing-bones show that rapid to and fro
(*i.e.*, up and down) movement of the wing would involve a larger
amount of exertion than in a wing with hollow bones as in most
modern birds. It is not so much the extra weight that would be
impedimental as the extra inertia. The narrowness of the body and
smallness of the sternum indicate that the air-cavities of the abdomen,
even if present, were smaller and less effective for respiration than in
modern birds. I may be permitted here to make a remark as to the
use of abdominal air-sacs in birds, or I may be suspected of having
fallen into an old error of supposing that they materially diminish the
weight of the bird. Flight involves, perhaps a greater, *i.e.*, more
rapid, consumption of energy than any other form of locomotion, and
all powerful fliers, whether birds or insects, are provided with
respiratory organs of enormous effectiveness. In birds, instead of air
being only pumped into the bronchial tubes and the rest being left to
diffusion, the air is drawn tight through the lungs into the abdominal
and other air-sacs, so that the highly vascular lung with its venous
blood supply is brought into more direct relation with the "tidal air"
than in mammals, while the "residual air," which is not changed at
each double respiratory movement, rests, not as in mammals, in the
lung itself, or at least not chiefly so, but in the air-sacs outside the
lungs. The respiratory organ proper is thus brought into direct

relation with air capable of far more rapid renewal than in the
mammalian lung. In other words, birds breathe the tidal air, while
mammals breathe the residual air. *Archæopteryx*, however, shows no
sign of the possession of large air-sacs, or of that large expanded
sternum which, in modern birds of flight, insures the rapid change of
the air by the same muscular movements as are involved in flight
itself. While in all powerful fliers, both birds and insects, every
movement of the wing insures a change of the air in the respiratory
organ itself (and not merely in the passages leading to that organ),
the form and structure of *Archæopteryx* forbid us to believe that such
an adaptation existed in it.

Conclusions as to the size and efficiency of the heart might be
drawn, but they are not only obvious but also perhaps a little risky;
so I will leave them.

Nobody, except the constructors of flying-machines, seems to be
ignorant of the fact that all powerful fliers have a wing-area which is
very large in proportion to the area of the immoveable aëroplanes.
Birds do not nowadays rely upon immoveable aëroplanes at all. The
wings serve all the functions of both propulsion and sustentation, and
the tail when used at all is used only for steering; while the vastly
superior flight of insects is effected in the absence of any special
steering apparatus, the wings themselves serving alike for propulsion,
sustentation, and steering.

In *Archæopteryx*, then, we find an animal as yet not evolved beyond
the aëroplane phase of flight; a phase characterised by the use of
large aëroplanes, which, while offering considerable resistance to flight,
take no part in the propulsion. It may further be pointed out that
among flying things, whether birds, mammals, insects, or flying
machines, the most efficient are all far broader from side to side than
they are long. Lilienthal, alone among men who seek to fly, seems to
have appreciated the real bearing of this fact. The further mechanical
consideration of the point, even as applied to *Archæopteryx*, would,
however, take us too far. It is sufficient for my present purpose to
have pointed out wide differences of structure both as to the immediate organs of flight and also as to the propelling muscles and the
respiratory organs on which powerful flight depends, between, on the
one hand, *Archæopteryx*, and, on the other hand, all other flying

animals.   I will only add an argument for the benefit of biologists
who may be unable to grasp the significance of mechanical considerations.   The oldest known bird—*Archæopteryx*—was constructed, so
far as the sustentatory apparatus of flight was concerned, on a principle which has been superseded in all modern birds, and which may
therefore be safely pronounced, even on purely biological grounds, to
be inferior in effectiveness to the principle of construction which has
superseded it.

Archæopteryx was, therefore, not a very good flier.   How good it
was as a flier, I dare not guess; though the perfect, but rather small,
wings strongly suggest that, in spite of the impediments I have mentioned (free digits, heavy head and neck, large aëroplanes offering
resistance to rapid movement, small muscles, heavy non-pneumatic
bones, and deficient respiratory apparatus), *Archæopteryx* could fly
better than the competing pterodactyles.

Since the days of *Archæopteryx*, its descendants, or the descendants
of its near relatives, have been evolved into the modern bird, which,
in its more perfect forms, is a perfect biped and a powerful flier : and
a brief enumeration of some of the chief changes involved in that
evolution will throw an additional light upon my contention as to the
comparatively small power of flight, and as to the quadrupedal locomotion of *Archæopteryx*.   The shoulder has shifted backwards, the
trunk has become shorter, bringing both elbow and knee-joints nearer
the centre of gravity ; thereby rendering balance in walking or standing
independent of the aid of the fore-limb.   The fore-limb, in accordance
with this release from one of its functions, has lost the digits I, II,
and III on which that function depended, and has thus been reduced
in inertia, while the pelvis has become widened to allow for the backward displacement of the abdominal viscera, and the tibia has acquired
a cnemial crest of much greater size, enabling the hind-limb to support
the weight of the whole body, with the knee flexed and placed in such
position that the centre of gravity is behind it.   The aëroplanes of the
tibiæ have been abolished, and that of the tail shortened, lightened,
and reduced to a steering apparatus pure and simple.   The wings
have gained in size ; their inertia has been reduced by the development
of air-cavities in the wing-bones ; more powerful muscles have been
evolved for the movement of the wings, and this has involved the

development of a pectoral crest and of a large sternum. The reduction in weight of the head (especially jaws) has aided in the backward shifting of the centre of gravity. The development of the large sternum has facilitated the evolution of a respiratory apparatus of more efficient type, and the shortening of the whole body and concentration of its weight near one point, aided by the stiffening of the back—often by ankylosis of the vertebræ and always by the increase in length of the functional "sacrum" has rendered the support of the hinder part of the body by cumbrous aëroplanes unnecessary.

IV.—ANSWERS TO CRITICS.

Many and various arguments have been urged against my view as to the homology of the digits of the wing of an ordinary bird, as well as against my interpretation of the *photograph of Archæopteryx* and other contentions contained in an article on sources of error, which was published in "Natural Science" (Vol. III. p. 275).

Mr. Pycraft asks (p. 445), Where did my supposed but unseen digits IV and V articulate in *Archæopteryx?* I have answered this already in the preceding pages of this article; but there is good reason for answering it again. The digits did exist, and their metacarpals are conspicuous in the London specimen, and they articulated to that large rounded carpal bone which is shown at the distal end of the radius and ulna of the right wing in Fig. 2, p. 116.

Mr. Virgil L. Leighton, in No. iii. of "Tuft's College Studies," says (p. 71), "That there is developed a fourth digit in the avian manus is beyond question [!], and the fact that this comes upon the ulnar side of the three permanent fingers is sufficient to invalidate the nomenclature III, IV, and V of Hurst." The "plausibility" of my view is, moreover, founded on ignorance, and "no one without a theory to support would regard it [*i.e.*, what I called *os pisiforme* in Nat. Sci., Vol. III., p. 279] other than a digit." Mr. Leighton, of course, knows that greater men than myself regard the *os pisiforme* itself as the remnant of a digit; but even if he did not, the alleged resemblance of it to a digit in a certain stage in the development of a certain bird (*Sterna*) would not in the least affect my conclusions. It is not difficult to name a few mammals in which things occur much more like digits beyond the normal five than the embryonic rudiment he refers

to. (N.B.—Rudiment = beginning, germ, thing as yet undeveloped,
" Anlage.")

Only one really strong and rational objection to my views has come
to my notice.   It was urged in conversation by Dr. Jaekel, of Berlin,
and subsequently by Professor Dames.   They urge the principle, well-
known among palæontologists, that in all fossils we must expect the
evidence of things below the surface to appear on the surface ; that all
the bones of a single specimen usually lie nearly in a single plane.
The bones of *Archæopteryx* in the Berlin specimen certainly do all lie
almost in a single plane.   There is, however, no evidence in that
specimen of the existence of the supposed digits IV and V.   What I
formerly took for a shadow in the photograph (*see* 12, Plate XV.) is
merely a yellow stain on the slab (but not without significance), and
the slender digits I, II, and III are not crushed or distorted as if by
underlying bones ; the tibial quills, as I have called them, of the right
leg do unmistakably show a displacement or distortion where they lie
over the left knee ; vertrebral and ventral ribs lie in juxtaposition, and
so even do the ribs of right and left sides.

This, however, not only ceases to be an objection, but actually comes
to support my view when we enquire more closely into it.   Many fossils
are distorted much as they would be by enormous vertical pressure :
they are flattened much as wax models of them would be by being put
in a hydraulic press and squeezed.   But *Archæopteryx* is not so distorted
The skull is not flattened : its cavity was empty, and not either filled
with matrix or abolished by the collapse of its walls.   The left femur
does *not* lie in the same plane as the right, but so inclined that its head
is now deeply embedded in the slab, while that of the right lies well
above the general surface of the slab.   The digits do not lie all in
one plane ; in the right hand the second lies over the third crossing
it without displacement and without deformation.   On the other hand,
the first metacarpal lies *lower* than the second, and not above it as it
should in the natural position of the parts.   The second metacarpal
and proximal phalanges lie, not in the same plane as the feathers, but
well above that plane.   Whatever flattening there might have been it
could not have *lifted* the second digit well above the level of the
proximal ends of the quills it was supposed to have supported.   The
bones of a dead bird, which has fallen on the bottom of a stream, or

lake, or sea, will naturally fall to the level of the underlying deposit
as fast as the decay of the soft parts allows them to do so. The weight
of overlying deposits may even flatten out ribs and bring them all to
one level. It may even crush the bones where they lie one over the
other—I am not sure that this has not occurred in the left hand—
but it has not, in this case, crushed the skull, or the phalanges of the
third digit of the right hand, or the pelvis; and it has not brought
the proximal ends of the two femoral bones into juxtaposition. Had
it brought the digits IV and V to the same level as I, II, and III, we
should have seen them, and these latter might well have been crushed
by them; but the perfect preservation of these digits, even where they
cross each other, and the fact that they do lie, especially the second,
well above the level of the feather-surface, shows that there has been
in this case little, if any, deformation by pressure of overlying strata;
hence the absence of all trace of digits IV and V on the surface. If
any such deformation had occurred it would have brought II and III
to the level of the feather-surface. No matter, therefore, what pressure
there may have been, the fact that those digits II and III lie now
above that surface shows that they did originally lie above it, and not
below it as all views except my own demand.

C. HERBERT HURST.

LITERATURE REFERRED TO IN THE FOREGOING ARTICLE.

OWEN.—*Phil. Trans.*, 1863.

DAMES.—Ueber Archaeopteryx. Berlin, 1884.

PYCRAFT.—*Natural Science.* Vol. 5, p. 350 and p. 437.

LEIGHTON.—*Tufts College Studies*, No. III.

HURST.—*Natural Science.* Vol. 3, p. 275 ("Errors")

　　　"　　　"　　　"　　Vol. 6, p. 112, p. 180, and p. 244.

(The present article is in large part reprinted from NATURAL SCIENCE. *Vol. 6.)*

EXPLANATION OF PLATES.
PLATE XIV.

Photograph of left wing of *Archæopteryx*, from the specimen in the Natural
History Museum, Berlin (*two-thirds nat. size*).

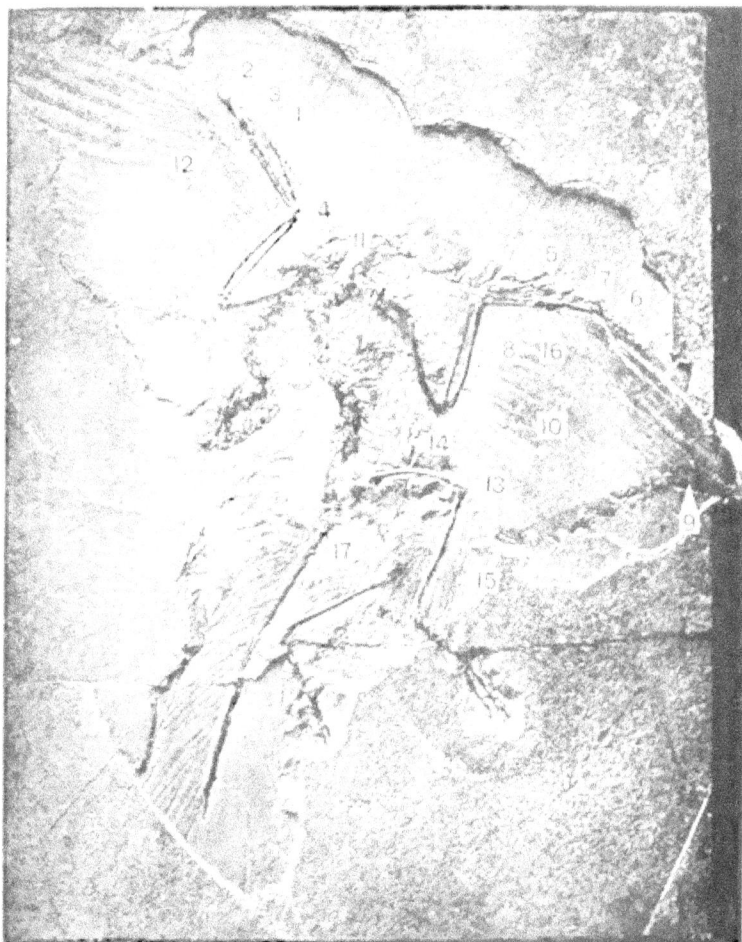

PLATE XVI.

PLATE XV.

A photograph of the Berlin specimen of *Archæopteryx* taken before the skull and certain other parts were as fully exposed as they now are. (Scale 5 : 17).

1. First digit of left manus. 2. Second ditto. 3. Third ditto. 4. Radiale of left carpus. 5, 6, 7. First, second, and third digits of right manus. 8. Region of right wing which in the specimen lies lower than the visible bones of the hand and lower than the region marked 10. 9. Primary quills of right wing. 11. Small portion of matrix lying upon proximal end of humerus. (See reference to this in the text.) 12. Yellow stain resembling, in the photograph, a shadow. 13. Cnemial crest. 14. "Abdominal" or ventral ribs. 15. Feathers of crural aëroplane. 16. Joint between second and first phalanges of third digit. 17. Left femur.

The specimen was illuminated at the time of photographing by light falling upon it from above and in front and slightly to the left.

PLATE XVI.

ARCHÆOPTERYX.

From a photograph, one-quarter natural size, of the specimen in the British Museum (Natural History), taken by kind permission of Dr. Henry Woodward, F.R.S., Keeper of the Geological Department.